Sedimentology and Sedimentary Basins

From Turbulence to Tectonics

Sedimentology and Sedimentary Basins

From Turbulence to Tectonics

Mike Leeder
Formerly at the School of Earth Sciences,
University of Leeds
Now at the School of Environmental Sciences,
University of East Anglia

Blackwell
Science

© 1999 by
Blackwell Science Ltd
Editorial Offices:
Osney Mead, Oxford OX2 0EL
25 John Street, London WC1N 2BL
23 Ainslie Place, Edinburgh EH3 6AJ
350 Main Street, Malden
    MA 02148 5018, USA
54 University Street, Carlton
    Victoria 3053, Australia
10, rue Casimir Delavigne
    75006 Paris, France

Other Editorial Offices:
Blackwell Wissenschafts-Verlag GmbH
Kurfürstendamm 57
10707 Berlin, Germany

Blackwell Science KK
MG Kodenmacho Building
7–10 Kodenmacho Nihombashi
Chuo-ku, Tokyo 104, Japan

First published 1999

Set by Graphicraft Limited, Hong Kong
Printed and bound in Great Britain
by MPG Books Ltd, Bodmin,
Cornwall

A catalogue record for this title
is available from the British Library

ISBN 0–632–04976–6

Library of Congress
Cataloging-in-publication Data

Leeder, M.R. (Mike R.)
    Sedimentology and
    sedimentary basins:
    from turbulence to tectonics /
    Michael Robert Leeder.
        p.   cm.
    ISBN 0–632–04976–6
    1. Sedimentology.
    I. Title.
    QE471.L375   1999
    552'.5—dc21   98–53090
                        CIP

DISTRIBUTORS

Marston Book Services Ltd
PO Box 269
Abingdon, Oxon OX14 4YN
(*Orders*: Tel: 01235 465500
            Fax: 01235 465555)

USA
Blackwell Science, Inc.
Commerce Place
350 Main Street
Malden, MA 02148 5018
(*Orders*: Tel: 800 759 6102
            781 388 8250
            Fax: 781 388 8255)

Canada
Login Brothers Book Company
324 Saulteaux Crescent
Winnipeg, Manitoba R3J 3T2
(*Orders*: Tel: 204 837–2987)

Australia
Blackwell Science Pty Ltd
54 University Street
Carlton, Victoria 3053
(*Orders*: Tel: 3 9347 0300
            Fax: 3 9347 5001)

For further information on
Blackwell Science, visit our website:
www.blackwell-science.com

For Marta Pèrez-Arlucea

Time was away and somewhere else.

Louis MacNeice,
'Meeting Point', *Collected Poems*, Faber

Contents

Colour plates fall between pp. 288 and 289

Part 1
Introduction

1 Sedimentology in the earth sciences

. . . the soil which has kept breaking away from the high lands during these ages and these disasters, forms no pile of sediment worth mentioning, as in other regions, but keeps sliding away ceaselessly and disappearing in the deep.

Plato, *Critias*, Vol. 9, Loeb Classical Library

1.1 Introduction: sedimentology and earth cycling

Sedimentology, the study of transport and deposition of sediment, provides information for geologists to interpret sedimentary rocks. Sediment is derived from breakdown of rocks and minerals at and near the Earth's surface. Protons and electrons from the atmosphere and hydrosphere, along with the physical action of salt and ice crystallization, act together as weathering agents to reduce the relief produced by plate tectonics. But classical physics tells us that matter and energy can be neither created nor destroyed. To the sedimentologist, this principle means that global balances must occur between atmosphere, continent, ocean and solid earth. This aspect, recognized long ago in its essential aspects by James Hutton, is the very stuff of sedimentology, cutting across stilted traditional disciplines of geology, physics and chemistry. Thus all elements involved in earth processes undergo recycling and must be accounted for in sediment budgets. Recycling processes have gone on for $> 3.3 \times 10^9$ yr (Erikkson *et al.*, 1994; Nijman *et al.*, 1998), thoroughly mixing but also stratifying the Earth. More recently the activities of *Homo sapiens* have directly impinged on these natural processes, particularly those of direct concern to sedimentology: hence current concerns about acidification, 'greenhouse' warming, soil erosion and wholesale global pollution. These issues have many implications for sedimentologists since the Earth's atmosphere has almost certainly evolved through geological time, a point to which we shall refer on future occasions.

The water cycle (Fig. 1.1) links atmosphere and climate with the rock cycle of plate tectonics. Without surface runoff, erosion and sediment transport rates would be very much reduced and totally dependent upon simple falling and toppling due to gravity, extra-planetary impacts and so on. It is easy to make out a case for the water cycle being the most fundamental to affect the Earth's surface: it lies at the roots of just about everything. Unfortunately the cycle as a whole is not often investigated, but instead only parts are emphasized by geologists, meteorologists, oceanographers and civil engineers in the light of their particular specialism. The cycle influences climate through the high heat storage and heat transfer capacities of water and water vapour, particularly when phase changes from gas to liquid to solid are involved and latent heat is transferred. It is also not generally realized that water vapour is far more important as a 'greenhouse' gas than CO_2 and may be expected to have a serious positive feedback effect upon planetary warming. A final aspect of the water cycle concerns the change from seawater to rainwater to ice and back again. We note here that this process fractionates molecules of water which have 'light' oxygen, $H_2{}^{16}O$, compared to those which have 'heavy' oxygen, $H_2{}^{18}O$, the latter a tiny minority. This is because more energy is required to evaporate the heavier water from the sea, and so water vapour has proportionately more light water and so, therefore, have the continental ice-sheets. Thus the waxing and waning of ice-sheets sends a tiny but measurable signal immediately (in geological terms) into the whole oceanic reservoir because of the relatively short ocean mixing times involved. The signal is then recorded in the shells of oceanic calcareous plankton.

The mean residence time of atmospheric water is small because of the disparity between the vast oceanic reservoir and the water content of the atmosphere in relation to annual evapotranspiration and

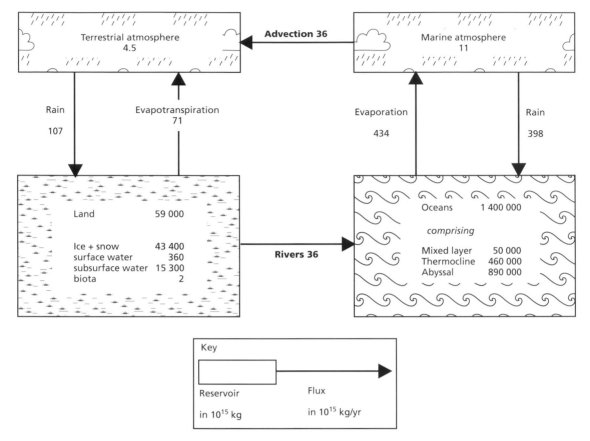

Fig. 1.1 The Earth's water cycle and major reservoirs. Mean residence intervals for particular reservoirs may be calculated by dividing the resident mass by the annual outward flux rate. The accuracy of some components of the cycle is poor, resulting in a closure error of about 2. Continents and oceans exchange water, with precipitation exceeding evaporation over land, the excess returning as runoff: it is this runoff that is responsible for the transport of sediment from land to the oceans. (After Chahine, 1992, and references therein.)

precipitation. The whole water content is recycled 33 times per year, giving a mean residence of only a week or so. The mean residence time of the oceans as a whole is some 3000 yr, but this is not the same for all of the oceanic water masses, values for the surface layers being only days or weeks. The fast regimes of change in near-surface water have an important role in modulating the slow regimes of change in deep oceans and glacier ice.

The fastest rates of element recycling and reaction at and near the Earth's surface are associated with the biosphere, which fixes major elements such as oxygen, carbon (Figs 1.2 & 1.3), hydrogen, nitrogen and sulphur. Each of these elements also has stable isotopes, whose fractionation during elemental flux from tissue to atmosphere or hydrosphere and back again leaves

tell-tale signs behind in the reaction products. We can make use of these biogeochemical cycles in assessing the dynamics of earth recycling.

The oxygen cycle (Walker & Drever, 1988) concerns us especially in sedimentology because of the role of the element in oxidative weathering of organic matter and pre-existing silicate minerals. Although life-supporting, O_2 is at the same time highly corrosive to rocks and minerals that formed under conditions of elevated temperature and pressure inside the Earth, where *free* oxygen is very rare. Also of importance is its control on the precipitation of minerals from surface waters or seawater, particularly variable-valency transition metals like iron and manganese. In addition, the presence of oxygen-deficient seawater enables the preservation by burial of abundant organic

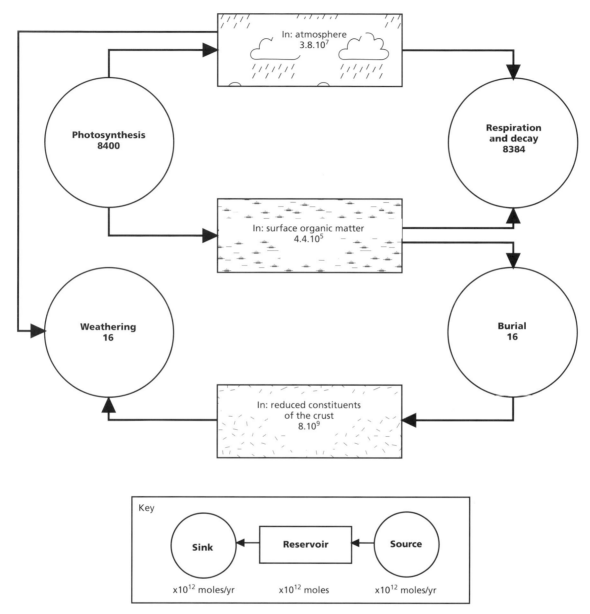

Fig. 1.2 Oxygen budgets, fluxes and biogeochemical cycles. All numbers are expressed in 10^{12} mol of oxygen (or capacity to combine with oxygen) per year. The chief point to notice at this stage is the large reservoir of atmospheric oxygen and the small but important role played by the burial and weathering of organic matter in the sedimentary rock cycle. (Adapted from data in Walker, 1980; Drever *et al.*, 1988.)

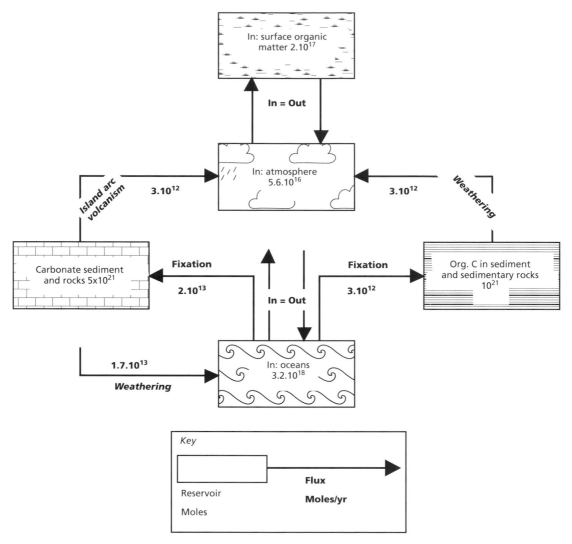

Fig. 1.3 The pre-industrial global carbon budget and cycles for times longer than 10^4 yr. Note the huge amount of carbon in fossil fuels and limestones ($> 10^7$ Gt) compared to the atmosphere and living surface biota. The largest reservoir outside of fossil fuels and limestones is the oceanic dissolved pool of CO_2, HCO_3^- and CO_3^{2-}. Exchanges between atmosphere and ocean are by gas transfer, and between the biota and atmosphere by photosynthesis and respiration. Siegenthaler and Sarmiento (1993) estimate the current anthropogenic carbon flux to the atmosphere as about 6×10^{14} mol/yr. Most comes from fossil fuel combustion: about 50% of this is taken up by the atmosphere and about 25% by the surface ocean (although the estimate is controversial). The fate of the remaining 25% or so is unknown but may be taken up by the terrestrial biosphere and/or by the 'oceanic carbon pump'. The latter would export it into cold oceanic bottom waters for subsequent medium-term (10^3–10^4 yr) re-export to the surface ocean in upwelling areas. (Adapted from data in Walker, 1977; Drever *et al.*, 1988.)

matter. It seems well established from the work of popularizing scientists like Lovelock (1979) and Westbroek (1991) that the biosphere has created the atmosphere as we experience it, probably early on in the Earth's history (though exactly *when* is still highly controversial). Atmosphere composition reflects a balance between oxidative weathering, photosynthesis, respiration and burial of organic carbon (Fig. 1.2). The current level of atmospheric ('free') oxygen is 21% or so. Before the presence of abundant oxygen in

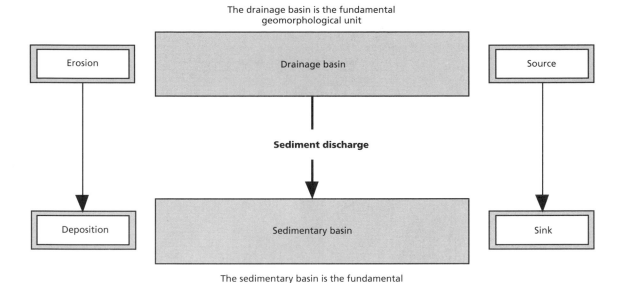

Fig. 1.4 Simple sketch to contrast drainage basins and sedimentary basins, stressing that together the two concepts unify all geology and geomorphology, especially sedimentology.

the atmosphere and its dissolved fraction in the oceans, we must surmise that the mineralogy of sediments must have been quite different. Such notions have led to an upsurge of interest in recent years in sedimentary rocks of pre-Proterozoic age (roughly pre-2.5 Ga).

The carbon cycle is of central importance to our subject because it involves not only the organic carbon components of sediments and sedimentary rocks, like hydrocarbons, but also the volumetrically predominant $CaCO_3$ precipitated and secreted in the oceans and in limestones. The cycle involves CO_2 fixed from the atmosphere by reduction during photosynthesis into the organic individual. The carbon is then oxidized back to CO_2, continuously during respiration and finally upon death by the omnipresent aerobic bacteria. This is where the O_2 and CO_2 cycles interact and the recycling concept becomes seamless. Environmental and geological circumstances, such as anaerobic water conditions, burial and tectonic subsidence, conspire to preserve some portion of the dead organic matter. Even more carbon is fixed because the solubility of CO_2 in water (and here the water and carbon cycles converge) encourages reactions of carbonate and bicarbonate ions with dissolved calcium ions to form organic and inorganic $CaCO_3$ precipitates.

1.2 Erosional drainage basins and depositional sedimentary basins

There is an extremely close relationship between the sciences of sedimentology and geomorphology. Nowhere is this more obvious than in the inter-relationship between drainage basins (catchments) and sedimentary basins. Drainage basins, like sedimentary basins, may be defined on many scales, from tiny catchments developed on fault scarps to continental-scale examples like the Mississippi or Amazon. Chorley, in a memorable phrase, has called the drainage basin 'the fundamental unit of geomorphology'. In the same way, many geologists would view the sedimentary basin as the fundamental unit of geology. The two types of basin are complementary (Fig. 1.4): one is generally the site of net erosion, providing a flux of sediment to the other. Long-term preservation of some of the flux occurs due to subsidence brought about by tectonics. Clearly, the rate of erosion in the drainage basin upstream of the sedimentary basin will control the flux of sediment across the sedimentary basin.

1.3 Global sediment discharge and earth recycling: the rock cycle

Estimates of past and present sediment discharge from the continents have clear relevance to the debate about earth cycling and the history of continental

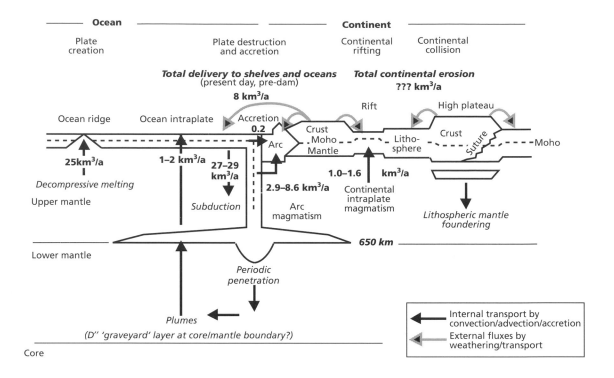

Fig. 1.5 The magnitude of internal and external discharges to and from the Earth's crust and mantle. The accuracy of the estimates of discharge is poor in some cases, preventing a firm conclusion as to whether the crust is growing or declining in volume or at steady state. (Data from Leeder, 1996, and sources cited therein.)

crustal growth. Mass-balance estimates for continental growth rely largely upon comparison of rates of sediment subduction with magmatic additions over tens or hundreds of million years (Reymer & Schubert, 1984; von Heune & Scholl, 1991; Taylor & McLennan, 1995). Time variations in sediment supply have been neglected in these estimates, as has the partitioning of deposited sediment between oceanic and continental crust. Estimates of pre-dam (*c.* AD 1900) Holocene sediment discharges seem to converge around 8 km³/yr, or about 2.1×10^{10} t/yr. Thus Milliman and Syvitski (1992) compute the pre-dam discharge at about 7.6 km³/yr. Summerfield and Hulton (1994) gathered data for major river systems draining some 35% of the continental land area. Scaling up their discharge of about 2.6 km³/yr gives a global figure of about 8.3 km³/yr.

Figure 1.5 indicates the magnitude of the annual global sediment discharge in comparison with published estimates of other magmatic and structural contributions to the continental crust. But is present-day sediment discharge a guide to past discharges of interest in longer-term balancing studies? The answer commonly considers that we should be more interested in pre-farming (i.e. pre-Neolithic) or pre-Quaternary discharges for such comparisons. The former quantity is difficult to estimate. Concerning the partition of sediment discharge between continental and oceanic basins, it is clear that the majority of sediment is deposited in sedimentary basins and continental margins formed by stretching and thermal subsidence of the continental crust. Some proportion escapes the shelf and is deposited directly on ocean crust in submarine fans, as thermohaline 'drifts' and by mass-wasting processes from the steep slopes of passive continental margins. It is this reservoir of sediments that is eventually transported on the subduction 'conveyor belt' when the margin turns destructive. Present-day major submarine fans comprise a volume of some 8×10^6 km³ (see Pickering *et al.*, 1989). Their ages are those of the major rivers supplying them and nowhere (apart from the Niger and Mississippi) does this exceed about 40 Ma. This

Part 2
Origin and Types of Sediment Grains

gives a mean sediment discharge to the oceanic crust of about 0.2 km³/yr, not including the contribution from contour drifts and mass wasting. These latter can scarcely be more than the fan contribution but their estimation awaits systematic study. We may roughly estimate a likely maximum figure for the total sediment discharge to the ocean crust of about 0.5 km³/yr.

Today, much of the discharge of sediment to the ocean crust (> 90%) occurs along passive oceanic margins and, although deposited on ocean crust, cannot begin to be partly digested into the mantle before the margins become destructive. Thus it seems that continental area must be periodically reduced by erosion and then enlarged by accretion as plate frameworks are adjusted on a timescale of several hundred million years. It is possible that such long-term unsteadiness may help to explain apparent correlation (Taylor & McLennan, 1995) of crustal growth with supercontinent accretion and dispersal cycles. However, given the magnitude of the oceanic sediment discharge compared with the total crustal volume (~8×10^9 km³), the effect seems to be small (about 1% over 100 Myr).

Recycling processes have created the Earth, but how has the Earth's lithosphere, hydrosphere and atmosphere changed through time?

- When is the first evidence for surface water?
- How has oceanic composition evolved?
- Is the continental crust growing, shrinking or at steady state?
- At what rate did photosynthesis create atmospheric oxygen?
- How did rocks weather before O_2 and CO_2 dominated the atmosphere?

- What effect has the evolution of plant life had upon weathering, erosion and sediment supply rates through geological time?
- Has the gravitational constant and hence the transport rate of sediment changed through geological time?

We shall return to the more sedimentologically relevant of these questions at various points during this book.

1.4 Comparative interplanetary sedimentology

In order to appreciate properly the nature of physical and chemical processes acting upon the surface of the Earth, let us look briefly at some other planetary bodies in our Solar System. Study of Table 1.1 will reveal a number of major differences between Earth, Mars and Venus. Earth is unique in its occurrence of surface water, high oxygen/low carbon dioxide atmosphere, low surface temperatures combined with moderate surface pressure and active plate interactions. Earth is a sedimentologist's paradise, though extraterrestrial volcanism and tectonics are perhaps more fresh and obvious on Venus. Exciting results from Mars since the late 1970s indicate that previously in its history there was much surface runoff, and as I write (January 1998) the first probe for many years to our Moon goes out to seek possible permafrost in polar lunar craters.

One reason for including this section is to illustrate the important principle that any explanation for sediment transport on Earth should also be general

Table 1.1 Surface conditions on Solar System planets.

	Venus	Earth	Mars
Temperature (K)	750	288	218
Pressure (mbar)	90 000	1013	7
Gravity (m/s²)	8.88	9.81	3.72
Atmosphere	CO_2 96%	N_2 77%	CO_2 95%
	N_2 3.5%	O_2 21%	N_2 2.7%
		H_2 1%	Ar 1.6%
Adiabatic lapse rate (deg/km)	10.7	9.8	4.5
Rotation rate (d)	−243	1	1.03
Solar flux (W/m²)	2610	1370	590
Albedo	0.7–0.8	0.39	0.15
Surface water	no?	yes	previously
Plate tectonics	previously	yes	previously

enough to be applicable to any other planet. To illustrate this, the founder of physical sedimentology, R.A. Bagnold, tells us in his autobiography (Bagnold, 1991) that at the age of 80 or so he was called upon to consult with NASA concerning the likely intensity of Martian sandstorms. NASA were concerned about the possibility of damage to their landed spacecraft by the impact of sediment transported by intense Martian winds, as suggested by spaceprobe images. Bagnold was able to apply his well-established theory of wind-blown sediment transport directly from Earth to Mars because it was founded upon the general principles of physics, rather than just by a correlation based upon observations. This lesson is part of a more general one concerning the superiority of approaches that are based upon the equivalence in different flow systems of a few fundamental principles, for example the importance of dimensionless quantities like the Reynolds and Froude numbers, whose delights we shall sample later in this book.

1.5 Practical sedimentology

Sedimentology is an eminently practical subject, with large numbers of individuals employed in the subject worldwide. There are a number of economic and environmental reasons for this. The natural hydrocarbons (oil, gas and coal) occur almost entirely within sedimentary rocks. Sedimentary rocks are also important aquifers and they host a wide variety of metallic and nonmetallic ores, including much of the world's copper, gold, iron and uranium, just to mention a few. The detection and exploitation of all these resources requires increasingly sophisticated and novel methods of study. For example it is now possible to make quite detailed predictions for the origin of hydrocarbon and water reservoirs from a combination of three-dimensional (3D) seismic reflection profiles, electric log signatures and downhole probes, and all this without necessarily having actually seen any sedimentary rock! Such models can of course be quite wrong, but the point is that taking core is highly expensive and can only be done in carefully chosen situations. It is thus obvious that the industrial sedimentologist should have a good basic grounding in process sedimentology, as well as an appreciation of the basic sciences and a wide range of other geosciences. The industrial problems of tomorrow will always be best solved by imaginative and creative science in

the spirit of the times, rather than by well-worn formulaic manipulation or consultation of encyclopaedias. It is this living, complex, interdisciplinary nature of the subject that the author tries to stress in this book.

In addition to resource-led applications, sedimentologists play a great role at the cutting edge of all endeavours in environmental engineering. By this I mean the management, safeguarding and training of natural depositional environments like coastlines, rivers, lakes and so on. For example, there are growing problems worldwide with coastal erosion, river flooding and dam siltation that directly involve the skills of process sedimentologists. Here again it is essential that the traditional engineering approach of 'designing to safe limits' is joined by an appreciation of fluid-dynamics first principles and an appreciation of natural complexities and environmental sensitivity. For example, engineers often do not appreciate the interrelations between parts of the natural environment, and when asked, say, to design a scheme for local river flood alleviation, have often just dug out and straightened a natural stream regardless of the consequences to the channel, the surrounding floodplains and wetlands. This is not a new trend, for Aldo Leopold raged against it in the USA in the 1920s, 1930s and 1940s (see collected essays in Leopold, 1991), and there are many interesting accounts of more modern depredations (e.g. Purseglove, 1989). Similar comments may be made about the management of hillslopes and the problems of soil erosion. Perhaps one of the most intriguing examples of our approach to sedimentological change concerns the natural avulsion of the Mississippi into the Atchafalaya basin and the consequent long legal struggle (involving professional sedimentologists and engineers) to define exactly when a lake becomes a river (Steinberg, 1995).

In all of the many sedimentary environments to be discussed in this book, there are similar practical issues arising from their reaction to tectonic, sea-level and climate change. These issues are producing some of the best applied scientific research of the present time. Of course in many of these areas the sedimentologist must work very closely with engineers, geomorphologists, hydrologists, oceanographers and others. But it is only through an appreciation of the complexity and diversity of natural sedimentary environments that true understanding will come. Here is a flavour of just some of the major issues:

- How will a new dam in the headwaters of a major river affect the floodplain and channel downstream, viz. will the channel aggrade or incise?
- Will global warming lead to greater or lesser soil erosion in areas already prone to its worst effects?
- How and when will managed or 'clear-cut' deforestation have a detrimental effect upon slope stability and sediment erosion?
- How will coastal wetlands and beaches react to sea-level rise?
- How quickly can the oceans mop up the excess CO_2 produced by human polluting activities?

1.6 A brief history of sedimentology

This is *not* the place for a long account of sedimentology down the ages. The Egyptians were acutely knowledgeable about the timing and magnitude of the processes whereby precious water and silt were dumped on Nile floodplains. The Phoenicians designed harbour moles and jetties specifically to avoid problems of siltation. The Greeks recorded progradation of their coastlines and erosion of deep old soils after deforestation. They also had 'River Gods' whose changing fluviatile moods towards their equally capricious human subjects feature most prominently in Homer's *Iliad*. The Minoans could accurately depict fluvial landscapes (see back cover).

The interpretation of sedimentary rocks played a major role in the development of geology, from Hutton's demolition of Werner and the role of erosion and sedimentation in his great scheme of global recycling, to Lyell's pioneering analysis of sediment-ary structures. In the 1840s and 1850s, Sorby developed Lyell's results and laid the foundations of sediment-ary petrology, pioneering the analysis of bedforms and sedimentary structures from a hydraulic viewpoint. The elucidation by German, Swiss and Austrian geologists of nappe structures in the 1920s and 1930s relied on the use of sedimentary structures for 'way-up' evidence. Bagnold's desert explorations in the 1930s and subsequent work on the physics of sediment transport opened up new eras in the interpretation of bedforms and sedimentary structures.

The need for industrial insights into petroleum reservoirs led to a huge expansion of research into modern sedimentary environments, particularly deltaic and shallow marine carbonate environments by US and UK oil companies in the 1950s and 1960s. At the same time the engineering and military needs of wartime and 'Cold War' USA led to intensive study of, first, river behaviour and coastal processes and, later, the subsea exploration of shelves and oceans. The general theory of plate tectonics evolved from observations made by Wegener and du Toit on the similarity of Permo-Carboniferous glacial sedimentary deposits in what we now know as the Gondwana supercontinent. More recently, the Deep Sea Drilling Project and its successor, the Ocean Drilling Project, have cored hundreds of kilometres of sediment from the ocean floors of the world, shedding much light on palaeoceanography and palaeoclimates from Mesozoic times onwards. Nowadays much sedimentological work is done by geochemists and geophysicists who find the complex problems of sedimentology of very fundamental interest to the working of the Earth 'machine'.

2 Water–rock interactions: chemical and physical breakdown of catchment bedrock to soil and clastic sediment grains

Few ken to whom this muckle monument stands,
Some general or admiral I've nae doot,
On the hill-top whaur weather lang syne
Has blotted its inscribed palaver oot.

Hugh MacDiarmid, 'The Monument', *Complete Poems*, Vol. 1, Carcanet

2.1 Introduction

Clastic sedimentary rocks are usually quite different in their mineralogy from the igneous and metamorphic rocks that comprise their 'primary' sourcelands. We may illustrate this by noting that, although feldspar is the commonest mineral in the Earth's crust (about 60% of the total), quartz is usually predominant in clastic sediments and sedimentary rocks. But the principle of conservation of mass tell us that, for all elements present in the exposed crust and released by weathering, exactly the same levels of abundance must occur in sediments (assuming constant ocean water composition). Thus the average *chemical* composition of *all* sediments is roughly that of a granodiorite, approximately the composition of the average whole crust.

It is traditional to divide rock weathering into physical and chemical components, but in reality the two are inextricably interlinked. Chemical and biochemical weathering usually involve aqueous reactions at mineral surfaces in the unsaturated (vadose) zone that lies close to the local Earth's surface, usually within 100 m or so. Living and dead vegetation plays an important, sometimes dominant, modulating role. A soil profile results whose characteristics depend upon climate and rock type. Water is the chief reactant and plays a dual role since it also transports away both dissolved and solid weathering products. An outline of the near-surface terrestrial hydrological cycle is given in Fig. 2.1.

Aqueous reactions have a strong biochemical component since they are aided by dissolved atmospheric and soil-generated gases, dilute acids and organic ligands. The reactions are complex since silicate minerals are involved, with their many constituent anions

and cations; also the amount of water and dissolved ions varies in both time and space. Four main mechanisms contribute to chemical weathering: dissolution, oxidation, hydrolysis and acid hydrolysis.

Physical weathering involves the application of *in situ* and *ex situ* stresses to rock and mineral discontinuities in the unsaturated zone, causing fragmentation. Stresses are set up due to unloading, gravity, wind shear, salt crystallization from groundwaters, freeze–thaw and differential thermal expansion. The action of physical weathering by itself does not produce a soil layer—an additional *in situ* organic contribution is required to do that.

It is the combined effects of chemical and physical weathering that produce a weathered profile. This comprises:

- surface soil with significant organic contributions;
- chemically altered saprolite that has significantly changed chemically;
- chemically unaltered but often physically fragmented (exfoliated) bedrock.

Interfaces between these layers are in a state of slow downward motion as the landscape reduces. In fact, landscape dating by cosmogenic isotopes and other means reveals that a steady-state system often exists, with the material mass removed by erosion being replaced by an equal volume made available for further decomposition.

Weathering involves:

- bond breaking, physically by cracking and chemically by solution;
- broken bond (danglers) adoption by ionic predators;
- electron removal from the easily stripped transition metals like Fe and Mn.

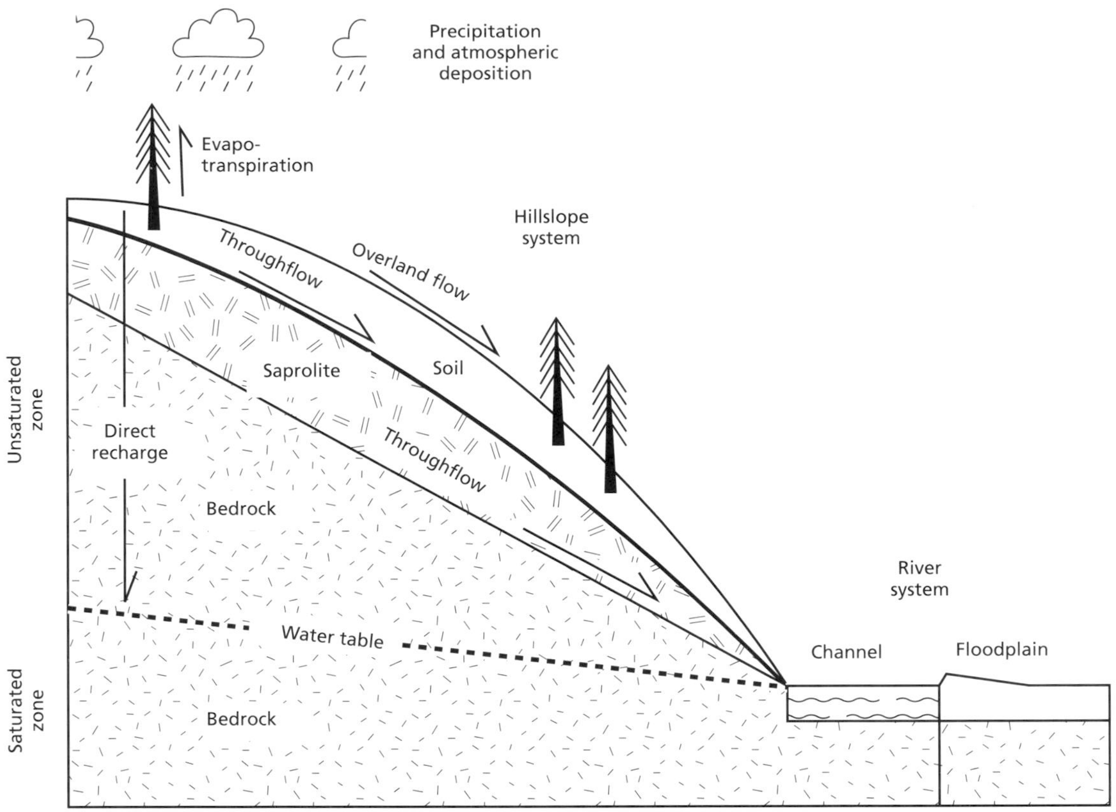

Fig. 2.1 The main components of the terrestrial hydrological cycle.

Weathering acts on:
- mineral aggregates (rocks!), not simple laboratory chemicals from a bottle;
- imperfect crystals with surface and lattice defects.

Weathering depends on:
- H_2O throughput;
- H_2O dissociation into the reactive H^+ and OH^- ions;
- CO_2 concentration via atmospheric and soil processes;
- temperature, via the results of the Arrhenius principle.

2.2 Natural waters as proton donors: pH, acid hydrolysis and limestone weathering

Water molecules separate into H^+ (protons in this context; actually as hydrated protons, H_3O^+, the oxonium ion, in reality) and OH^- (hydroxyl) ions at all temperatures. Protons are responsible for the acidity of an aqueous solution and enable us first to consider weathering in terms of proton donors and budgets. The molar concentration of H^+ ions in dilute solutions is so small and variable that acidity is expressed on a logarithmic scale, as the negative logarithm of the free H^+ concentration. This is the pH, with units of mol/kg (note that the higher the pH, the lower the proton concentration!). At room temperature there are only 10^{-7} mol/kg of protons (with an equal number of OH^-) in pure water, and this is used as a reference value for a neutral pH of 7. Larger values of pH are termed *alkaline* and smaller values *acidic*. These conditions are possible because H^+ or OH^- ions may be provided by other reactions and reactants in aqueous solutions.

No natural waters are pure—all contain greater or lesser quantities of ions in solution (solutes) in addition to the ubiquitous H^+ ions. We would naturally like to be able to understand what concentrations of

More information on the properties and role of water in rock weathering

Earth is unique in its abundance of water and water vapour. It is easy to take the stuff for granted, yet the deceptively simple molecule H_2O has remarkable properties of great import for rock and mineral weathering. When placed between the plates of a charged capacitor, the molecules orientate themselves with the positive hydrogens towards the negative plate and the negative oxygens towards the positive plate. The effect is similar to the orientation of a magnet in a magnetic field—water molecules possess poles just like a magnet and similarly act as electric dipoles. The strength of the dipole moment is determined by the product of the magnitude of the charges and the distance between them.

Polar water molecules result from asymmetric covalent bonding where the strongly electronegative (electron-loving) oxygen atom takes a majority share of available electrons. This gives oxygen a partial negative charge at the expense of a partial positive charge on the less electronegative hydrogen atoms. The chemical bonds between the oxygen and the two hydrogen atoms arise from the overlap of the p-electron clouds of the oxygen atom with the electron clouds of the hydrogen atoms. The bonds should, theoretically, be at 90° to one another and consist of a molecular orbital occupied by a pair of electrons. In practice, because of repulsion between the hydrogen atoms, the bond angles meet at an angle of just over 105°, the O–H bond length being 9.8×10^{-8} mm. When the water molecules come together *en masse*, they interact loosely by hydrogen bonding because of the polar nature of the H_2O covalent bonds. They form into tetrahedral groups of four by this process (Fig. B2.1).

Many of the distinctive properties and reactive tendencies of water are explained by its molecular structure.

1 It is a highly effective solvent for ionically bonded compounds (NaCl is a familiar example) because of its polar nature. In this process of dissolution the positive and negative ends of the water molecule attach themselves to the compound's negative and positive surface ions respectively, neutralizing their charges and enabling mechanical agitation to separate the constituent ions. Once in solution the ions are surrounded by water molecules whose combined electrostatic attraction may partly balance the ion's potential energy, thus stabilizing it by hydration, an ion–dipole interaction. The degree of stabilization by hydration is determined by ionic potential, i.e. the ratio between the electrostatic charge and ionic radius (see further below).

2 The tetrahedral groupings of water molecules cause such properties as high surface tension and capillarity. High surface tension causes water to have a high saturation effect in porous materials, penetrating tiny cracks and crevices.

3 Decreasing temperature is accompanied by an increase in density as a result of contraction caused by decreasing molecular thermal agitation. This effect is opposed by the production of more and more hydrogen bonds. The minimum density of fresh water is reached at 4 °C but this

continued on p. 18

More information on the properties and role of water in rock weathering [*continued*]

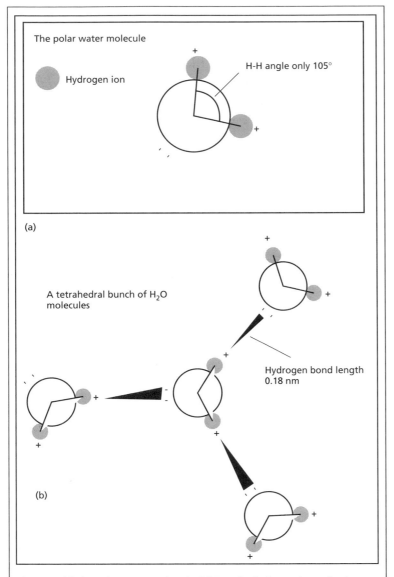

Fig. B2.1 (a) The polar water molecule. (b) Tetrahedral groupings of polar water molecules.

temperature is depressed by increasing pressure and salinity. Expansion dominates below 4 °C as the tetrahedral molecular groups also begin to be taken up into hexagonal ice structures and so the density decreases. Expansion continues until −22 °C at which ice achieves its minimum density and maximum expansive pressure. Hexagonal ice crystals have a maximum growth rate normal to the basal plane, so that ice whisker crystals growing in microcrevices of rocks (as noted above, water is highly penetrative because of its high surface tension) can exert high boundary stresses.

reactants are produced from particular reactions. But what about the state of saturation of a solution? How can we tell whether an aqueous solution like rainwater or stream-water will dissolve rock-forming minerals (i.e. a state of undersaturation exists) or precipitate minerals (state of oversaturation exists)? Take a surface weathering example. A kilogram of natural stream-water is found by analysis to contain 2×10^{-3} g of $CaCO_3$ in solution. A quantity such as this is known as the ion activity product (IAP) or concentration. Since the relative molar mass of $CaCO_3$ is $40 + 12 + (3 \times 16) = 100$, we have what is known as the activity (a) of the solution as $0.002/100 = 2 \times 10^{-5}$. Such a solution is extremely undersaturated since laboratory experiments would show that it could dissolve appreciably more $CaCO_3$. In fact at 25 °C the water could hold 2.91×10^{-2} g $CaCO_3$, an order of magnitude more than our natural sample. The water is thus said to be undersaturated and capable of dissolving more $CaCO_3$ locally as it travels around.

Since many natural aqueous solutions come into contact with the atmosphere, we must also express the concentration of a gaseous component (CO_2 for example) in a solution. This is done by noting that any gas exerts a pressure (P) across an interface with fluid and that this will be in direct proportion to the mole fraction (x_i) of component i in the gas. Thus we speak of the partial pressure (p_i) of component i as:

$$p_i = x_i P \qquad (2.1)$$

Many natural weathering and sediment-forming reactions are reversible in that they can move either 'forwards' or 'backwards' depending upon local conditions of availability of ions, dissolved H^+ and so on. Thus in pure water (no H^+ or OH^- ions) calcite dissociates by the reversible reaction:

$$CaCO_3 \rightleftharpoons Ca^{2+} + CO_3^{2-} \qquad (2.2)$$

A state of equilibrium exists for every such reaction in which the rate left to right (forwards) is exactly balanced by the rate right to left (backwards). At this equilibrium the activity of the reaction products on either side of the equation are balanced, so we may define a constant term K (for given temperature) as:

$$K = \frac{a_{Ca^{2+}} a_{CO_3^{2-}}}{a_{CaCO_3}} \qquad (2.3)$$

K is called the equilibrium constant for any particular reaction at a specified temperature. For this case $CaCO_3$ is a pure solid and for all solids activity $a = 1$. Thus:

$$K = a_{Ca^{2+}} a_{CO_3^{2-}} \qquad (2.4)$$

If we disturb the system, say by adding calcium ions, then the reaction will work to neutralize the change, preserving the value of K, in this case by moving into reverse to produce (precipitate) solid calcium carbonate. Generally for most salts, whose dissolution in water is an endothermic (needing heat energy) process, an increase of temperature increases the value of K and hence the solubility increases. This is not so for gases, for when temperature and/or pressure increase solubility decreases. This makes the calcium carbonate weathering system in natural waters particularly complex, since dissolution is not controlled simply by the dissociation reaction above, but by reactions involving dissolved CO_2 (see eqn 2.5) whose source may be atmospheric or respiratory. In such cases p_{CO_2} is the critical quantity that controls pH and the rate of reaction.

Perhaps the most important source of H^+ ions is provided by atmospheric CO_2 dissolved in rain and by other humic and bacterially produced acids. Also, abundant biogenic CO_2 is produced by respiration of microorganisms in soil water. Thus we must examine the following series of interlinked reactions for the combination of water and carbon dioxide and the two-stage ionization of carbonic acid:

$$H_2O + CO_2 \rightleftharpoons H_2CO_3 \rightleftharpoons H^+ + HCO_3^- \rightleftharpoons 2H^+ + CO_3^{2-} \qquad (2.5)$$

The dissociation of carbonic acid is strongly pH-dependent, the equilibrium pH of pure water in the H_2O–CO_2 system of eqn (2.5) being about 5.7 at 25 °C. The equilibrium constants for the above reactions and for the whole carbonate system at various temperatures are given in Table 2.1. It can be seen that:

1 CO_2 is, perhaps surprisingly, less soluble at higher temperatures.
2 The first dissociation of carbonic acid (K_1) produces most hydrogen ions for acid reaction.

These conclusions have many sedimentological applications, to which we shall return from time to time.

It has been found by field measurements and laboratory simulations (Keller & Wood, 1993) that p_{CO_2}

Table 2.1 Various equilibrium constants* (K) for the carbonate system of eqn (2.5).

$T\,(^\circ C)$	K_{CO_2}	K_1	K_2	$K_{calcite}$	$K_{aragonite}$
5	$10^{-1.19}$	$10^{-6.52}$	$10^{-10.55}$	$10^{-8.39}$	$10^{-8.24}$
25	$10^{-1.47}$	$10^{-6.35}$	$10^{-10.33}$	$10^{-8.48}$	$10^{-8.34}$
60	$10^{-1.78}$	$10^{-6.29}$	$10^{-10.14}$	$10^{-8.76}$	$10^{-8.64}$
90	$10^{-1.94}$	$10^{-6.38}$	$10^{-10.14}$	$10^{-9.12}$	$10^{-9.02}$

* The equilibrium constants are as follows:

$K_{CO_2} = a_{H_2CO_3}/p_{CO_2}$, $K_1 = a_{H^+}a_{HCO_3^-}/a_{H_2CO_3}$, $K_2 = a_{H^+}a_{CO_3^{2-}}/a_{HCO_3^-}$, $K_{calcite}$ and $K_{aragonite} = a_{Ca^{2+}}a_{CO_3^{2-}}$.

can reach large values (several times atmospheric, viz. $\sim 10^{-3.5}$ bar) in the subsoil vadose zone because of the small rates of diffusion of respiratory CO_2 to the surface. This contribution of respiratory CO_2 in soils decreases the pH very substantially below that of normal rainwater. Such enhanced values of pH due to soil CO_2 are found to persist even in deep aquifers, such as the Floridan aquifer (Plummer, 1977), where the chemical conditions are important in controlling diagenetic reactions.

2.3 Metallic ions, electron transfer and E^h–pH diagrams

The loss of electrons from an element or ion is termed oxidation and leads to an increase in positive valency or a decrease in negative valency, and vice versa for reduction. For example, consider the oxidation of ferrous iron as might occur in water–mineral reactions involving silicate minerals, such as pyroxene, olivine, biotite or hornblende, or the ferrous disulphide, pyrite:

$$Fe^{2+} \rightleftharpoons Fe^{3+} + e^- \qquad (2.6)$$

Changes in oxidation state are important in chemical weathering and other sedimentary geochemical processes because they affect the solubility of metal species, particularly the oxides, hydroxides and sulphides of the transition-group metals. In the case of Fe, the Fe^{2+} (ferrous) form is a mild reducing agent because of the relative ease with which it surrenders its electron. The Fe^{3+} (ferric) form, by way of contrast, is a highly stable structural state because of its half-filled valence electron shell. The Fe^{2+}–O bond has a much lower site energy than the Fe^{3+}–O bond, and

therefore Fe^{2+} compounds are much more mobile (soluble). Highly important cycles of electron transfer occur between reduced and oxidized Fe and organic ligands (a ligand is an attaching compound, usually to a transition metal with its many vacant electron orbitals) at the boundary between oxygen-poor (anoxic) and oxygen-rich (oxic) zones. This produces gley-type soil horizons, blotched yellow-grey by mixtures of the two iron species. Dissolved oxygen in surface waters is the most important natural oxidizing agent on account of its very high electronegativity (second only to fluorine in its attraction for electrons).

It is possible to measure the oxidation–reduction potential (redox potential) by noting the potential difference (E) produced between an immersed inert electrode, usually platinum, and a hydrogen electrode of known potential. This redox potential for reactions involving protons, commonly termed E^h by geochemists, is compared to the arbitrary value of 0.00 mV for hydrogen in the reaction:

$$2H^+ + 2e^- \rightleftharpoons H_2 \qquad (2.7)$$

at 25 °C and 1 atm pressure at a concentration of 1 mol/kg (pH = 0). Negative values of E^h indicate reducing conditions, and positive values oxidizing conditions, with respect to the arbitrary hydrogen scale. If we know the standard oxidation–reduction potential for a particular reaction from laboratory measurements, and we possess a field measurement of E^h from a particular weathering zone, then it is possible to predict the type of dissolved oxidation state for a particular ion. For example, the standard potential of the Fe^{2+}–Fe^{3+} couple of eqn (2.6) is 0.77 V. If, for example, our field measurement shows a reading of 0.5 V under acidic conditions (pH = 2), we would be able to predict that Fe^{2+} is the stable phase. However, the reactions that determine E^h are often very slow and the aqueous environment does not quickly come to equilibrium with the measuring electrodes. Thus field redox measurements generally give only semiquantitative information and tell us little about reaction rates or the attainment of equilibrium.

A particularly informative graph may be produced by plotting E^h against pH (Garrels & Christ, 1965). The area emphasized in Fig. 2.2 shows the usual limits of E^h and pH found in near-surface environments. The lower limit to pH of about 4 is produced by natural concentrations of CO_2 and organic acids dissolved in surface and soil waters. The upper limit to pH of

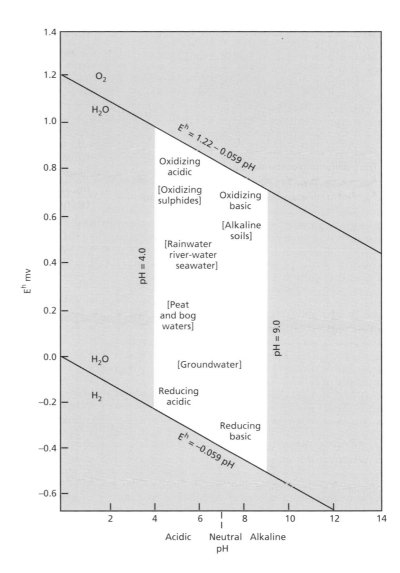

Fig. 2.2 An E^h–pH diagram to show the range expected in most earth-surface and near-surface aqueous environments.

about 9 is reached by waters in contact with carbonate rocks and still in contact with atmospheric CO_2. Local conditions may sometimes fall well outside the usual limits. Thus, oxidation of pyrite gives very acid conditions, as illustrated by the severe pollution caused by groundwaters issuing from many abandoned coalmines. Most elements in the weathering zone are oxidized. Exceptions exist in waterlogged soils or at the boundary between poorly and well-drained soils. Below the water table, conditions are oxygen-poor and anaerobic bacteria are abundant. Here the reaction in eqn (2.6) is reversed and insoluble ferric iron is reduced to soluble ferrous iron.

2.4 Behaviour of silicate minerals during chemical weathering: breakdown products and newly formed minerals

The flux of dissolved elements and altered minerals from continent to ocean is largely controlled by the processes of chemical weathering. The recharge of dissolved ions into rivers from the weathering zone (Fig. 2.3) is a function of the fluxes from atmospheric deposition, vegetation growth and respiration, and the net reaction of interstitial water with minerals in the unsaturated zone.

Chemical weathering thus plays a major role in the

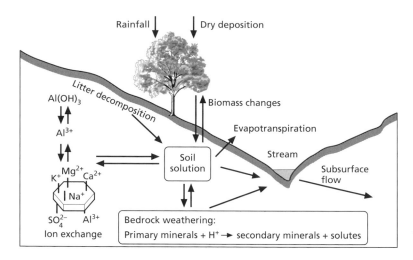

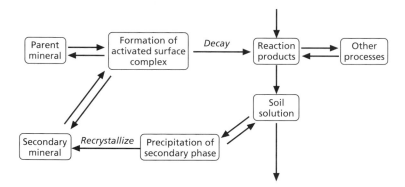

Fig. 2.3 Some of the processes affecting solute fluxes in catchments. (From Drever & Clow, 1995.)

Fig. 2.4 Flow diagram to show the characteristics of surface mineral reactions. (From Sverdrup & Warfinge, 1995.)

global hydrogeochemical cycle and therefore in environmental geochemistry. We have already noted that the range of abundant mineral types in clastic sedimentary rocks is usually very much more restricted than that in igneous and metamorphic rocks. Some minerals are clearly more stable than others in the weathering process. Some are newly formed by the weathering reactions, others suffer a net removal whilst yet others increase in abundance.

Chemical weathering reactions in the presence of natural waters include the rapid dissolution of ionically bonded minerals like halite and other evaporites and the acid attack on carbonate minerals. Both carbonates and evaporites usually dissolve congruently, i.e. the proportion of elements in solution is the same as the proportion in the former mineral(s). The decomposition of silicate minerals is predominantly by acid hydrolysis in which activated surface complexes play a major role (Fig. 2.4) and where small,

highly charged protons displace metallic cations in crystal silicate and oxide lattices. OH^- or HCO_3^- ions can then combine with the displaced cations to form solutions or local precipitates. Hydrolysis acts along lattice surfaces exposed by discontinuities such as joints, rock cleavage, crystal boundaries, mineral cleavage planes and crystal surface defect sites. Once a dissolved ion is liberated from a crystal lattice by hydrolysis, it may link up with surrounding water molecules and thus stay in solution as a hydrated ion, or, if the force of attraction with water is insufficient to compete with that between the water molecules themselves, the ion will be 'ignored' by the water and will precipitate. These alternatives are clearly of fundamental importance to the sedimentologist. Some elements will be exported from the weathering site, perhaps far away. Others will precipitate very locally, perhaps forming economic accumulations of the elements in question or of their compounds. Which kind

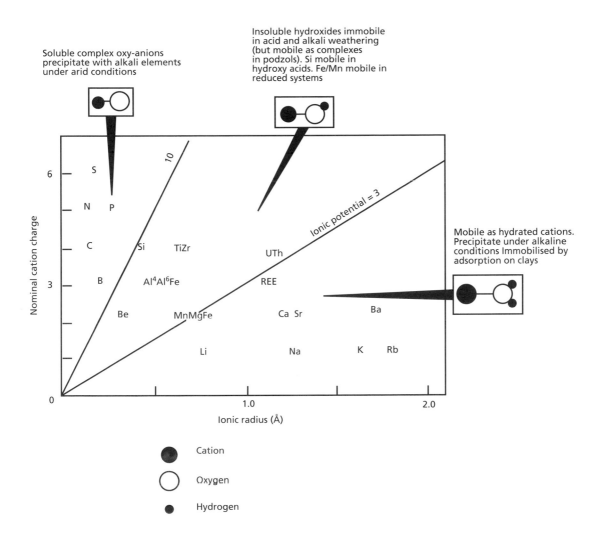

Fig. 2.5 The relationship between ionic charge and radius for various ions of importance in surface hydrogeochemical and weathering reactions.

of behaviour occurs in solution is determined by the ionic potential, which is the ratio of ionic charge to ionic radius (Fig. 2.5). Cations (metals) whose potential is less than 3 are easily hydrated and highly mobile; likewise for anions of potential > 12, which form soluble complexes. Ions with potentials between 3 and 12 are mostly precipitated as hydroxides (in the absence of protons) and are thus immobile.

The proton concentration of soil water usually gives rise to pH values in the range 5–9, in which silicon is more soluble than aluminium. Thus prolonged

chemical weathering will leach the Si from the soil profile, leaving behind a mixture of Al and Fe oxides and clays. These are lateritic (ferralitic) soils, common in well-drained tropical sites. By way of contrast, in highly acid soils Al and Fe are both leached to give a podzol with a characteristic light-coloured, silica-rich upper zone. The majority of silicate minerals, however, react very slowly and dissolve incongruently, giving other solid products in addition to dissolved ones. Reaction pathways vary according to local E_h–pH–temperature conditions. It has been calculated that silicate weathering accounts for some 45% of the total dissolved load to the world's oceans (Holland, 1978; Wollast & Mackenzie, 1983), with carbonate weathering accounting for 38% and evaporites 17%.

More information on the
thermodynamics of weathering
reactions

Chemical weathering is basically about breaking the atomic bonds or changing the electron configuration of atoms in pre-existing minerals making up the catchment bedrock. In order to break a bond or change an outer-shell electron's energy, energy must be exchanged and the energy budget balanced. In the first place, the likelihood of a particular reaction occurring may be decided by reference to the change in free energy of a reaction. The reaction free-energy change (ΔG) is the sum of the free energies of formation (ΔG_f^o) of all the reaction products minus those of the reactants. When calculated ΔG are negative, reactions will proceed spontaneously, although it must be said that the *rate* of reaction is not specified. The greater the negative value of ΔG, the more tendency there should be to react. In order to assess stability in this way we write specific weathering equations with the 'primary' igneous and metamorphic minerals as reactants. For example, consider Ca-feldspar (anorthite):

$$CaAl_2Si_2O_8 + 2H^+ + H_2O \rightarrow Al_2Si_2O_5(OH)_4 + Ca^{2+} \qquad (B2.1)$$

anorthite   in sol^n water   kaolinite   in sol^n

In the reaction for Ca-feldspar above, for example, ΔG is calculated as -23.9 kcal/mol (Curtis, 1976), indicating that anorthite will spontaneously react with hydrogen ions in aqueous solutions to form the clay mineral kaolinite plus calcium ions. The negative free-energy change indicates that the reaction will proceed spontaneously but the rate is *not* specified.

We may also assess the stability of minerals directly from solubility data for IAP and compare it with solubility products like k. When IAP/k = 1 then by definition ΔG is zero, and the various mineral phases are in equilibrium with the soil water. When IAP/k < 1 then ΔG is negative and the reactant mineral (Ca-feldspar in our example above) will dissolve. When IAP/k > 1 then ΔG is positive and the reactant mineral will precipitate. In all silicate weathering reactions it is found that IAP is directly dependent on H^+ activity, i.e. the reactions are strongly pH-dependent (Fig. B2.2).

Despite the elegance of thermodynamics, mineral–rock–water reactions in the oxidative and organic-rich weathering zone are *complicated*! There is no easy answer as to rates and mechanisms because of the large number of variables involved. We follow Lasaga (1995) in stating that any general rate law would generally and basically have to follow the Arrhenius exponential temperature law ($e^{-Ea/RT}$) developed in the text. But it would also have to include specific terms (discussed only qualitatively in this text) relating to the pre-exponential factors of mineral surface processes (k_0), the reactive area of the mineral including internal microporosity (A_{react}), the activity of hydronium ions (the H^+ ions, usually hydrated H_3O^+) raised to some power ($a_{H^+}^n$), the largely unknown effect of ionic strength ($f(I)$), the sum of catalytic and/or inhibitory effects of other ions like Al ($\sum a_i^n$), and, finally, the degree to

continued on p. 25

More information on the
thermodynamics of weathering
reactions [*continued*]

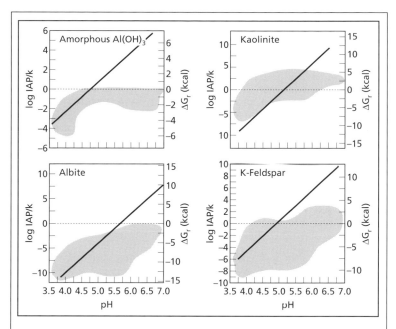

Fig. B2.2 The extent of dissolution, equilibrium and precipitation achieved by the reaction of aqueous solutions on various minerals at different pH under standard earth-surface conditions. (After White, 1995.) Shaded area indicates envelope of data points.

which the weathering reaction deviates from equilibrium ($\Delta G \neq 0$). All this is summarized in:

$$\text{rate} = k_0 A_{\text{react}} e^{-E_a/RT} a_{H^+}{}^n f(I) \Sigma a_i{}^n f(\Delta G) \tag{B2.2}$$

Given the complexities of this beast, it may come as a surprise to the reader that whole-catchment schemes based upon attempts to determine all these variables (see Sverdrup & Warfinge, 1995) computationally are in fact remarkably successful, a tribute to the skills of the field geochemist.

Quartz, muscovite and K-feldspar dominate amongst clastic mineral components that survive chemical weathering from weathered igneous and metamorphic terrains (Fig. 2.6). However, the abundance of these and other primary minerals is highly variable, depending upon a number of factors including climate and type of weathering, sourceland abundance, hardness, original grain size, rapidity of sedimentation, and so on (Suttner *et al.*, 1981; Nesbitt & Young, 1989). As usual, sedimentology cannot be reduced to mere chemistry or physics! There are also important newly formed minerals produced by weathering, which are considered here.

Quartz

Quartz (Dove, 1995; Hochella & Banfield, 1995) makes up 20% by volume of the exposed continental crust. Quartz crystals are spiral networks of linked silicon–oxygen tetrahedra; the lattice is thus extremely resistant to chemical attack (Dove, 1995; Hochella & Banfield, 1995). This is true over the acidic and neutral pH range but solubility increases in aqueous solutions at pH > 9. Below this value we have:

$$SiO_2(\text{quartz}) + 2H_2O \rightarrow H_4SiO_4(\text{aq}) \tag{2.8}$$

with the equilibrium constant, $K = 1 \times 10^{-4}$ at 25 °C.

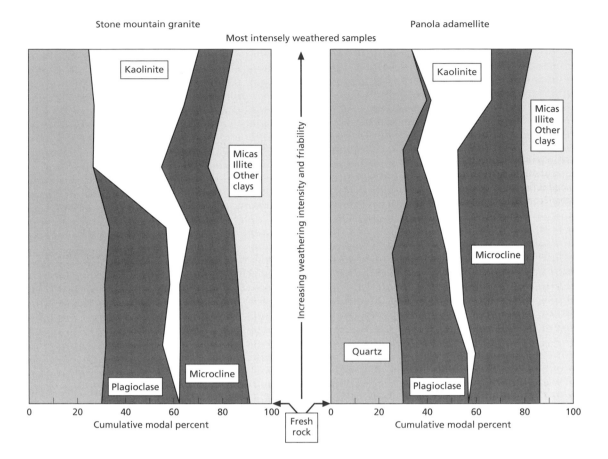

Stone mountain granite

Panola adamellite

Most intensely weathered samples

Increasing weathering intensity and friability

Kaolinite

Micas
Illite
Other
clays

Microcline

Plagioclase

Cumulative modal percent

Fresh
rock

Kaolinite

Micas
Illite
Other
clays

Microcline

Quartz

Plagioclase

Cumulative modal percent

Fig. 2.6 Mineralogical trends in weathering profiles developed on granite and adamellite. (After Nesbitt *et al.*, 1997.)

(We should note that at the same temperature K for *amorphous* silica is an order of magnitude higher at 2×10^{-3}.) At pH > 9 (Fig. 2.7) the H_4SiO_4 dissociates further (like the H_2CO_3 we discussed previously) into $H_3SiO_4^-$ and then finally, at pH values well outside those found in most weathering environments, $H_2SiO_4^{2-}$. At these highly alkaline pH values the crystal surfaces are subject to increasingly effective hydroxylation, particularly in the presence of alkali cations from NaCl and KCl, and the total dissolved equilibrium silica concentration increases rapidly.

Hydroxylation takes the form of so-called silanol surface groups by the reaction:

$$Si-O-Si + H_2O \rightarrow 2SiOH \quad (2.9)$$

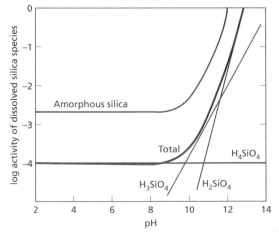

Fig. 2.7 Activities of dissolved silica species in equilibrium with quartz at 25 °C. The heavy curve is the total of activities. Top curve is the total of activities for equilibrium with amorphous silica. (After Drever, 1988.)

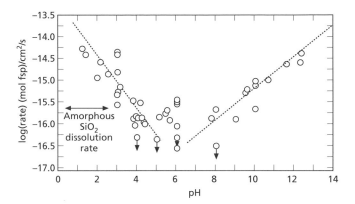

Fig. 2.8 Albite dissolution rate as a function of solution pH. Note the V-shaped curve common to many silicates. (From Blum & Stillings, 1995.)

Activation energies E_a for Si–O bond destruction are expected to be about 20% of the enthalpy of formation, ΔH_f^o, in this case about 90 kJ/mol. Silica polymerizes in highly silica-supersaturated solutions to form gels or sols. There is considerable evidence that the dissolution of quartz is greatly facilitated (perhaps by several orders of magnitude) by dissolved Al in the porewater environment. The main route by which quartz is made dissolvable, and also fragmentable, appears to be along microcracks and fractures.

Feldspar

Feldspar is the most abundant mineral in the Earth's crust, forming 60% of all igneous rocks for example. It is a framework aluminosilicate of Na, K and Ca in which Si–O and Al–O tetrahedra link to form an infinite 3D framework network with the alkali cations in the interstices. Feldspar weathering (Blum & Stillings, 1995) has been much studied because of its importance in soil acidification and global CO_2 fluxes. The rate of dissolution is an interesting V-shaped function of pH, in common with many silicates. As an example, the curve for albite is shown in Fig. 2.8.

At low pH the feldspar weathering reaction is simply a transformation by acid hydrolysis. Reaction rates are low at or about neutral pH, rising again because of hydroxylation at high pH. Here Si links (rather than the Al links under acid conditions) are thought to be susceptible to dissolution in the presence of base cations. The overall result of acid dissolution (worldwide probably the commonest form of soil weathering conditions) is the precipitation of kaolinite and the liberation of the alkali and alkaline-earth

elements in solution as hydrated ions, carbonate or bicarbonate ions with silica as byproduct. The main source of protons is dissolved CO_2, by the reaction in eqn (2.5). K-feldspar is also unstable under acid conditions and yields the products potassium carbonate, kaolinite and silica:

$$2KAlSi_3O_8 + H_2CO_3 + H_2O \qquad (2.10)$$
$$\rightarrow K_2CO_3 + Al_2Si_2O_5(OH)_4 + 4SiO_2$$

There are several other possible products of the reaction in eqn (2.10), notably the aluminium hydroxide gibbsite ($Al(OH)_3$), depending upon local conditions of pH, which control the nature of the dissolved Al species (see Drever, 1988). At very low pH (< 3) reactions such as that in eqn (2.10) give rise to preferential metallic cation leaching and the formation of a thick surface layer rich in Si. Such surface layers do not form at other pH values and the dissolution process is dominated not by diffusion through a surface layer but by direct bond breaking and scavenging of 'danglers' (broken bonds). Although there are a wide range of apparent activation energies and rate constants recorded in experiments, generally the observed field weathering order is

anorthite (Ca-plagioclase) > albite (Na-plagioclase) > orthoclase/microcline/sanidine (K-feldspars)

There is some evidence that the dissolution of K-feldspar is inhibited by the presence of Al in solution and certainly that mineral is the most persistent of feldspars in the weathering zone in many examples worldwide. Figure 2.9 shows the spectacular etch pits and channels formed by Holocene weathering of a perthitic alkali feldspar (Lee & Parsons, 1995, 1998).

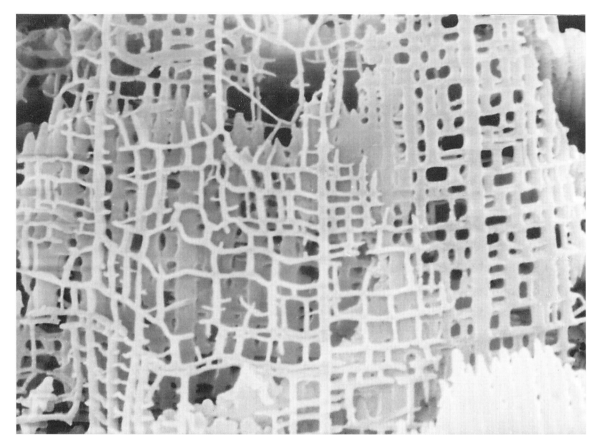

Fig. 2.9 SEM image of a resin cast of structurally weathered alkali feldspar from the Shap Granite, Cumbria, UK. The former grain surface is below the bottom of the image and the grain interior towards the top. The resin has penetrated into the interior of the grain along an intersecting network of etch pits that have formed by dissolution of highly strained feldspar surrounding edge dislocations. Field of view = 24 μm (Photo courtesy of Lee & Parsons, 1995, 1998.)

Micas

The distinctive thing about the micas (Nagy, 1995) is their sheet structure in which cations like Mg, K and Fe lie between layers of variously linked $(Si,Al)O_5$ tetrahedral and octahedral sheets. Cation substitutions give rise to negatively charged layers, which are neutralized by the cations in an interlayer. K^+ is the chief interlayer cation in muscovite, for example. Weathering reactions (Nagy, 1995) are chiefly controlled by hydration and hydroxylation at broken metal–oxygen bonds and the process is thought to occur progressively inwards from edge faces. Muscovite exhibits a typical V-shaped dissolution rate curve with respect to pH, with the lowest rates around pH 6 of about 10^{-13} mol/m^2 s. Biotite shows the same dissolution rate trend as muscovite at acid pH, but rates are usually much higher (5 times or more) because of the dissolution effects of oxidation upon Fe^{2+} ions.

Chain silicates

These comprise the pyroxene and amphibole groups in which silica tetrahedra are linked in either single (pyroxene) or double (amphibole) staggered chains by oxygen sharing. The chains are linked and held together strongly by metallic cations like Ca, Mg and

Fe linked to nonbridging apical oxygens. Amphiboles have in addition OH in the rings between opposite chains. Weathering (Brantley & Chen, 1995) is chiefly by surface reactions and acid hydrolysis by protonation, with rates of dissolution varying inversely with pH—but at vastly different rates for various members of the groups (see Brantley & Chen, 1995). But in addition, as with biotite mica, the occurrence of transition-group metals like Fe and Mn means that redox reactions with dissolved oxygen also occur, leading to generally faster rates of overall dissolution. Oxidation of liberated Fe^{2+} leads to surface crusting of amorphous Fe^{3+} precipitates, but it is not thought that these adhere significantly enough to lead to diffusion-controlled reactions.

Olivine

The olivine-group minerals have a simple structural arrangement of isolated SiO_4 tetrahedra linked by the divalent cations Mg^{2+} and Fe^{2+}. They comprise the end-members fayalite (Fe-rich) and forsterite (Mg-rich), with a complete solid solution series between. Fayalite and all varieties with any significant Fe (most of those in Nature) are highly susceptible to oxidative weathering and the formation of hydrated clay mineral and ferric crusts of goethite and haematite in humid saturated and arid unsaturated conditions, respectively (Hochella & Banfield, 1995).

Clay minerals

Clay minerals are one of the most important newly formed mineral groups in the weathering zone and they may form the bulk of the weathered residue. They may be easily transported to a depositional site where, in the absence of postdepositional changes, they give valuable information about weathering conditions (Fig. 2.10).

Kaolinite is formed under humid, acid weathering conditions from the alteration of feldspar-rich rocks by reactions such as that in eqn (2.10). Illite, a potassium aluminium hydrated silicate, is formed by weathering of feldspars and micas under alkaline weathering conditions where significant leaching of mobile cations such as potassium does not occur. Smectites are complex expandable sheet silicates with intracrystalline layers of water and exchangeable

cations. They form from the weathering of igneous rocks under alkaline conditions.

2.5 Acid rain and whole-catchment studies of chemical weathering

Acid hydrolysis reactions in soil zones with base cationic constituents (especially Ca^{2+}) usually lead to effective neutralization of rainwater, which in unpolluted areas has a pH of around 5.0. However, in many temperate zones with high rainfall and thin soils on Ca-poor substrates, the acidity is not neutralized, leading to formation of characteristic podzols (see section 2.10) with leached Al-poor surface horizons. Vastly increased industrial pollutants and emissions (CO_2, SO_2, NO) have accelerated these processes in many areas, including exports of gases to 'innocent' countries (e.g. Scandinavia), leading to extensive Al release in acidic waters and widespread environmental damage. The very steep increase of Al solubility in waters of low pH is illustrated by the data of Fig. 2.11. Assessments of acidification depend heavily upon estimations of cation fluxes, sources and sinks through the water–soil–bedrock system. Strontium-isotope studies have an important role here, as illustrated in studies in the northeast USA (Miller *et al.*, 1993). Very generally the solute-flux approach to catchment weathering seeks to draw up a balance sheet of all inputs and outputs to the catchment chemical system: it is basically a conservation-of-mass exercise. A simple expression (Sverdrup & Warfinge, 1995) states that

cation weathering rate = leaching + uptake − deposition − base saturation decrease

2.6 The rates and mechanisms of chemical weathering

An enormous amount of geochemical research has gone on in the past few years in an effort to understand and predict rates of chemical weathering (see White & Brantley, 1995). Much of this research has been environmentally funded or inspired because of the onset of acidification. Simply and generally, we may view the loss of mass, ΔM, in moles, resulting from chemical weathering over a time t, as a product

$$\Delta M = kSt \qquad (2.11)$$

where k is an overall rate constant ($mol/m^2/s$) and S

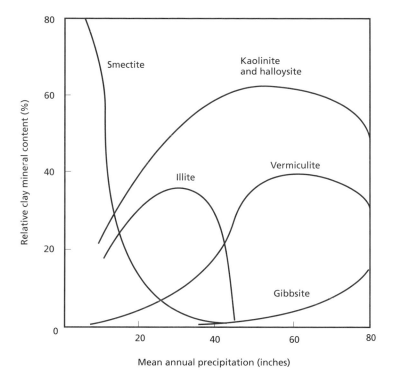

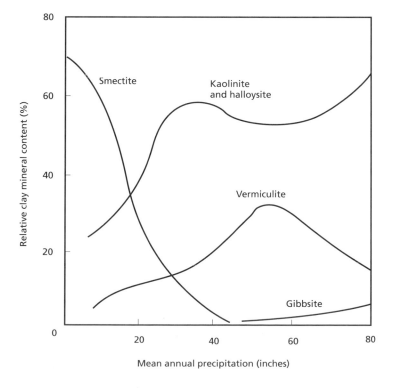

Fig. 2.10 Clay mineral composition of the surface layers of residual soils formed on acid and basic igneous rocks in California. (After Barshad, 1966.)

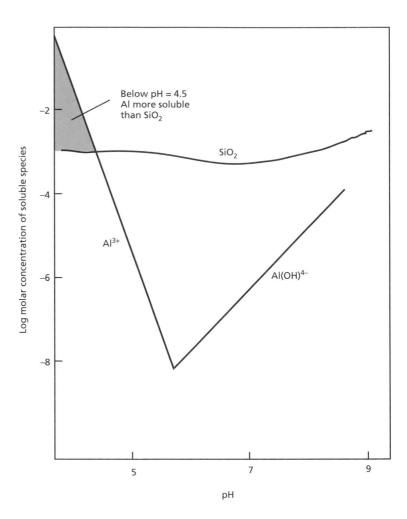

Below pH = 4.5
Al more soluble
than SiO_2

SiO_2

Al^{3+}

$Al(OH)^{4-}$

Log molar concentration of soluble species

pH

Fig. 2.11 The solubility of Al and Si as a function of pH. (After Raiswell *et al.*, 1980.)

is the total reactive area of mineral surface (m²). The temperature of weathering reactions is a key control over k. Like pure laboratory chemicals, natural minerals and the dilute acids that bathe them obey Arrhenius's law. This states that k increases exponentially with increasing temperature according to:

$$k = A \; e^{-E_a/RT} \tag{2.12}$$

where A, the pre-exponential factor with units as k, and E_a, the activation energy with units kJ/mol, are known as the Arrhenius parameters. Activation energies for the main silicate minerals fall broadly in the range 30–90 kJ/mol. The law is easiest to understand kinetically by considering that any spontaneous reaction needs sufficiently energetic collisions between reactants to cause it to happen; an energy theshold

must be surmounted. In the case of our silicate mineral surfaces, the warmer the aqueous phase, the more energetic will the attacking protons become and therefore the faster the reaction becomes. As Fig. 2.12 shows, the effect of the exponential form is highly important for temperatures of weathering varying about 25-fold from polar to tropical regions.

The reactive area term in eqn (2.11) is important for it means that the rate of chemical weathering is also highly sensitive to processes of physical weathering, like glacial grinding, which 'prepare' surfaces for chemical attack. Surface areas are conventionally measured in weathering horizons by dosing with N_2 and determining adsorbed monolayer volume, but the technique cannot reveal the awesome complexity of the internal pores present in many minerals, particularly

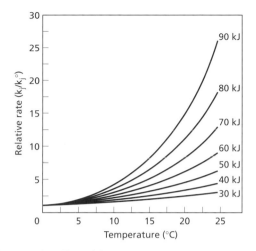

Fig. 2.12 The effect of the exponential Arrhenius temperature-driven term on weathering reaction rates for activation energies of reaction suitable for silicates (in the range 30–90 kJ/mol). (After White, 1995.)

feldspars with their turbid microporosity networks (Folk, 1955; Worden *et al.*, 1990; Walker *et al.*, 1995) that are entirely accessible to infiltrating waters.

Flow rate of water in the unsaturated zone is also important. Because much rainfall is recycled back to the atmosphere by evapotranspiration, it is also dependent on vegetation type and density. Optimal conditions for chemical weathering occur in forested highland areas of the humid tropics, where very thick soil and chemically altered bedrock zones have formed in the Holocene and where soil instability due to slope failure enables periodic exposure of fresh saprolite or bedrock. Lack of physical weathering and erosion leads to thick residual soils and a marked decrease in total export of ions. Such catchment slope conditions are known as transport-limited, for there is always an abundance of weathered product to transport.

The flow rate of water in the unsaturated zone works as follows. Most silicate reactions proceed to a point where the products have concentrations as predicted by the relevant equilibrium constant (for silicates these are *very* low). When the rates of forward and backward reactions become equal, no further weathering occurs. Only by removing the porewater can more weathering occur with 'new' waters. The rate of throughflow of water will basically be a function of rate of precipitation or runoff. This is why the

rate of chemical denudation usually shows a strong positive correlation with these variables. In order to account for the role of water concentration we may rewrite eqn (2.12) in the notional form:

$$k = f(P)A \; e^{-E_a/RT} \tag{2.13}$$

where $f(P)$ is a linear functional relationship involving some measure of effective monthly or yearly precipitation, P. The relationship is likely to be a complex one, if only because summer rains will be much more effective agents of chemical attack than winter rains. A form of eqn (2.13) successfully accounts for Si and Na fluxes from many catchments, but is unsuccessful for other alkali and alkaline-earth elements (White, 1995; Drever & Clow, 1995).

The role of water composition has been emphasized previously (in proton and electron transfers) and will not be repeated here, save to say that, when the E^h and pH of weathering solutions extracted from soils and weathered bedrock are determined, then they usually (semiarid, alkaline environments excepted) fall in the stability fields for the ubiquitous weathering product kaolinite.

Concerning the rate of dissolution of silicate minerals with time in natural environments, there is much experimental work (Furrer & Stumm, 1986; Schott & Berner, 1985) that suggests it is linear, controlled by surface reactions between mineral and aqueous phases (Fig. 2.13). Surface in this case refers to fractured and cleaved mineral faces. There is also sound evidence from electron microscopy studies that weathering can occur along uncracked and uncleaved areas of minerals, the process occurring along submicroscopic diffusion paths (Eggleton, 1986) at kink and step sites and other dislocations. The chemical attack is seen in the form of etch pits (Fig. 2.9). The rates of surface reactions are usually very slow compared to diffusion rates, and the concentration of products adjacent to the reacting mineral surface must be comparable to that of the weathering solution. In chemical language, a zero-order reaction rate law applies, so that:

$$r = \frac{dC}{dt} = ka \tag{2.14}$$

where r is the dissolution rate, seen to be directly linearly proportional to a, the surface area of mineral, and k, the reaction rate constant. Experimental dissolution (leach) rates for silicate minerals under acidic conditions are always low, being lowest for quartz

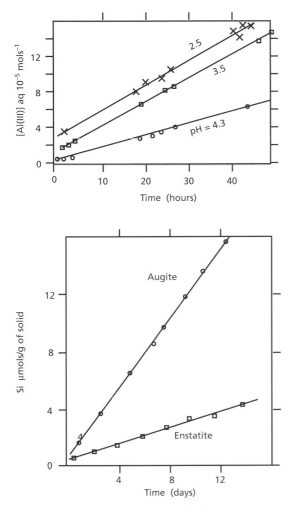

Fig. 2.13 Linear dissolution kinetics. (a) Al_2O_3; plotted is Al^{3+} released vs. time. (b) Frame silicates; plotted as Si^{4+} released vs. time for dissolution of etched pyroxenes at pH 6 and $T = 20–50$ °C. (After Schott & Berner, 1985; Furrer & Stumm, 1986; Stumm & Wollast, 1990.)

metal atoms and their linked oxygens to form precursor compounds. The nature and structure of these latter is the subject of much current research. It has been suggested that clays and other amorphous products may form directly in the surface layer of reaction, without perhaps a solution phase being passed through (Casey et al., 1993).

2 A slow detachment of metallic ions from the surface into solution. The latter is the rate-determining step and is controlled, amongst other things, by the probability of occurrence of sites at the mineral surface suitable for formation of the precursor compound. Experimental evidence (Holdren & Speyer, 1986) suggests that detachment reactions are nonstoichiometric, and that the order and ease of cation detachment is sensitive to pH. For example, under highly acidic conditions, aluminium is released from feldspar in preference to silicon (Muir et al., 1990; Casey & Bunker, 1991; Shotyk & Metson, 1994), with the formation of cation- and Al-deficient layers up to 1200 Å thick.

Although experimental studies indicate a linearity of dissolution rates, there is thus evidence from naturally weathered materials that rates decrease with time as a result of elemental exhaustion and/or the build-up of a thickening surface layer of weathered material that prevents aqueous access (Colman, 1981). However, a steady state may eventually be set up since some proportion of the weathered material can be removed by erosion. The local rate of chemical weathering will determine the thickness of weathered rinds that develop upon cobbles and pebbles in abandoned river courses or on terraces. With some sort of calibration to features of known age, it is then possible to use weathering rind thicknesses to determine further ages by extrapolation (Colman & Pierce, 1981; Knuepfer, 1988). The rate of chemical alteration of glacial-sourced debris in moraines and soils also shows a well-defined decrease with age (Taylor & Blum, 1995), probably because the finely ground mineral components dissolve first and easiest. Studies spanning 3.5×10^6 yr on soils atop dated basalt lava flows in Hawaii (humid subtropical forested environment) indicate that the easily weatherable base elements K, Mg and Ca, together with Si, are reduced to less than 10% of their initial values in the top 1 m of the soil profiles within 20 kyr (Fig. 2.15; Vitousek et al., 1997; Chadwick et al., 1999). More resistant Al, present in residual kaolinite, declines much more slowly, reaching such low levels only after 3 Myr.

(about 10^{-13} mol/m^2 s at pH < 6; Dove & Elston, 1992), compared with typical feldspar values up to 2–3 orders of magnitude higher.

How do surface reactions take place? We can envisage two steps (Stumm & Wollast, 1990; Fig. 2.14).

1 A fast attachment of hydration and protonation reactants like H_2O, H^+, OH^- and ligands for the transition metals like Fe and Mn to form surface species at the mineral–water interface. The reactants effectively polarize and weaken the bonds between

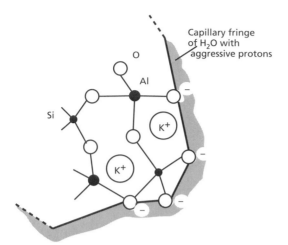

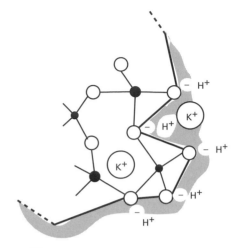

(a) Feldspar with broken surface bonds surrounded by capillary fringe of water

(b) Broken surface bonds are protonated; one ionic-bonded K^+ exchanged with H^+ from solution

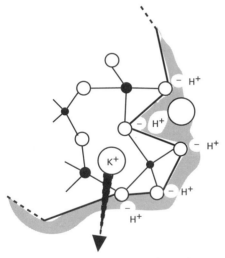

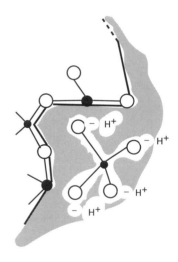

(c) Further exchange of K^+ for H^+ leads to complete protonation of the edge tetrahedron

(d) Edge silicon-oxygen tetrahedron removed in solution as H_4SiO_4

Fig. 2.14 Schematic view of highly simplified weathering mechanisms for the acid hydrolysis of a K-feldspar. (After Andrews *et al.*, 1996; see Blum & Stillings, 1995, for more detailed accounts of surface reactions.)

Tables 2.2 and 2.3 summarize work done on the relative rates of dissolution of common catchment rocks and minerals. It is noteworthy that the relative rates of mineral dissolution from laboratory experiments show far greater ranges than those determined from natural catchments underlain by single lithologies. Natural weathering is so much more complex, particularly in the flow and fluxes of the all-important aqueous phase.

2.7 A simple index of chemical alteration (CIA)

We have seen that the process of chemical weathering is complicated. Since most of us are not professional

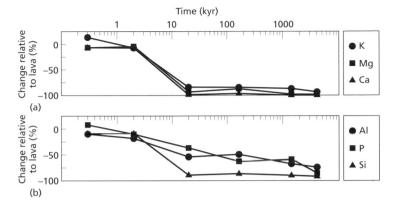

Fig. 2.15 The time changes of elemental proportions in soils developed on Hawaiian basalts of various ages. Note the most rapid rates of change in K, Mg, Ca and Si between 2 and 10 kyr. (From Vitousek et al., 1997; see also Chadwick et al., 1999.)

Table 2.2 Relative rates of CO_2 consumption and solute fluxes in runoff between monolithologic catchments. (From Amiotte Suchet & Probst, 1993.)

Rock type	Relative rate of CO_2 consumption	Relative rate of solute flux
Plutonic/metamorphic (granite, gneiss, schist)	1.0	1.0
Felsic volcanics (rhyolite, andesite, trachyte)	2.3	–
Basic volcanics (basalt)	5.0	–
Arenaceous sedimentary rocks	1.5	1.3
Argillaceous sedimentary rocks	6.6	2.5
Carbonate rocks	16.7	12.0
Evaporites	3.1	40–80

Table 2.3 Relative rates of dissolution of various minerals in laboratory experiments at pH 5 far from equilibrium. (From Drever & Clow, 1995, and sources cited therein; see also Sverdrup & Warfvinge, 1995, table 20.)

Mineral	Rate/rate for albite
Quartz	0.02
Muscovite (mica)	0.22
Biotite (mica)	0.6
Microcline (K-feldspar)	0.6
Sanidine (K-feldspar)	2
Albite (Na-plagioclase)	1
Bytownite (Na/Ca-plagioclase)	15
Enstatite (orthopyroxene)	57
Diopside (clinopyroxene)	85
Forsterite (Mg-olivine)	250
Dolomite	360 000
Calcite	6 000 000

geochemists it comes as a relief to discover a simple index that may be used to assess the degree of chemical weathering of bedrock. This is the chemical index of alteration (CIA) as defined by Nesbitt & Young (1982). We can simplify the weathering of the whole upper crust (to include 75% of cases) to that of a combination of feldspar and volcanic glass, involving only the commonest oxides Al_2O_3, CaO, Na_2O and K_2O. The CIA is given by:

$$CIA = \frac{Al_2O_3}{Al_2O_3 + CaO + Na_2O + K_2O} \times 100 \quad (2.15)$$

It has been found that $100 > |CIA| > 47$. The pristine upper crust has a mean CIA of 47, whilst the reader can verify that a value of 100 signifies wholesale removal of all alkaline earths. CIA values calculated for the suspended load of the major rivers of the world

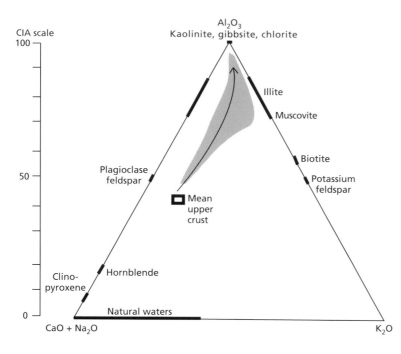

Fig. 2.16 Triangular diagram Al_2O_3–$(CaO + Na_2O)$–K_2O and chemical index of alteration (CIA) scale. (After McClennan, 1993.)

(McClennan, 1993) define a chemical weathering path (Fig. 2.16).

2.8 Vegetation, chemical weathering and the Precambrian controversy

It is tempting to regard chemical weathering as a test-tube process and as if our planetary surface was lifeless. But as we have mentioned several times, there is copious evidence that chemical weathering is modulated, perhaps even dominated, by vegetation. Higher plants intercept and evapotranspirate an appreciable proportion of rainfall, thus reducing potential infiltration and runoff. Not only do roots physically break up bedrock and saprolite, but associated symbiotic mycorrhizal fungi impregnate mineral surfaces with their filamentous hyphae, feeding the cations to the tree roots at astonishing rates: Jongmans *et al.* (1997) estimate that 150 m of pores in mineral grains are produced annually in 1 dm^3 of soil. Vegetated soil layers and litter also store precipitation and are a complicated chemical breeding ground for all manner of acid-producing reactions, chiefly those that generate soil CO_2 and other soil acids. Soil p_{CO_2} has a bearing on the recent debate concerning the likely efficacy of pre-Silurian chemical weathering, prior to the evolution of rooted vascular land plants. One school of

thought (Berner, 1992) holds that such weathering was minimal, whilst another (see Schwartzman & Volk, 1989; Keller & Wood, 1993) points to the key role of microbial respiration in the vadose zone and the very strong likelihood of a long pre-Silurian history for this type of H^+ input to chemical weathering and hence soil formation (see also review by Retallack, 1990). In addition it is likely that surface effects from lichen colonies played an important role in both rock penetration and stabilization, both processes increasing reactive area and weathering rates in pre-plant (cryptogamic) soils as compared with strictly abiotic conditions (Schwartzman, 1993). It has been estimated that rates of abiotic chemical weathering were perhaps up to 10^3 times less than at present. However, studies in inhospitable desert terrains (Campbell, 1979) and on young lava flows (Jackson & Keller, 1970; Berner, 1992) emphasize the importance of lichen and algal crusts, which almost certainly existed for much of later Precambrian (Proterozoic) times. The thallic 'roots' of the former also form a powerful biogeochemical microenvironment in which oxidation and disintegration processes are active, leading to formation of microsoil. The resulting weathered crusts ('cryptogamic' soils that may have a significant dust-blown contribution) are rich in ferric oxides and protected from physical

More information on chemical weathering: a buffer for global atmospheric and climatic change

A buffer reaction is one that uses up any increased quantity of a reactant produced by another reaction, leading to no change in pH or E_h conditions. So, we might ask, 'If *Homo sapiens* is altering the composition of the global atmosphere by emissions and deforestation, won't all the excess CO_2 just get used up by natural buffer reactions involving acid attack on limestone and calcium silicate minerals like plagioclase?' Consider first the role of carbonate weathering (Berner *et al.*, 1983; Berner & Lasaga, 1989). It is true that an increase in atmospheric CO_2 would increase global weathering of $CaCO_3$, but there is more to it than that (see Walker *et al.*, 1981; Walker & Drever, 1988). Consider the relevant equation:

$$CaCO_3 + CO_2 + H_2O \rightleftharpoons Ca^{2+} + 2HCO_3^- \qquad (B2.3)$$

Upon arrival in the oceans, Ca^{2+} and HCO_3^- can be stored in the short term (10^2–10^5 yr?) but it must eventually be extracted by organic and inorganic precipitation and CO_2 given off in the same proportion as provided to the atmosphere by the reverse reaction. A long-term steady state will thus be set up, controlled by the oceanic carbonate sink (see Plate 1). So the reaction cannot control increased CO_2 levels of the atmosphere in the long term. However, the insightful reader will note that control is possible *if* the oceanic biomass of planktonic primary producers can be decreased at the same time by another mechanism. This may happen naturally by nitrate removal (nitrate is a nutrient that determines primary productivity) brought about by denitrifying bacteria. There is evidence that during glacial periods the well-mixed oceans suffer less denitrification. The opposite is the case for interglacials, during which development of anoxic oceanic water masses encourages denitrification.

Of more use in global buffering on the long timescale ($> 10^5$ yr?) is the calcium silicate system (Berner *et al.*, 1983; Berner & Lasaga, 1989), which includes Ca-plagioclase and some chain silicates. We can write a simplified general global weathering equation (the Urey equation) as:

$$CaSiO_3 + 2CO_2 + 3H_2O \rightleftharpoons Ca^{2+} + 2HCO_3 + H_4SiO_4 \qquad (B2.4)$$

Here, for every two carbon atoms used up in weathering—forward reaction of eqn (B2.4)—only one will be returned to the atmosphere by the ocean carbonate sink by the reverse reaction of eqn (B2.3). It is now thought that the Urey equation is the chief feedback mechanism for the maintenance of a 'mean' global climate on a timescale of 10^6 yr or so (Walker *et al.*, 1981). It does this because, as CO_2 increases, there is increased global warming as a result of the greenhouse effects of that gas (see Chapter 13). But this warming will increase weathering reaction rates (the Arrhenius effect), thus using up CO_2 by negative feedback. It also works against any changes in the solar radiation constant, since increased radiation gives increased weathering temperatures, which uses up CO_2 and hence decreases the greenhouse effect.

erosion by stabilizing organic films and coats. It has been estimated that this microbially modulated weathering may have been only an order of magnitude less effective than higher-plant weathering (Schwartzman & Volk, 1989; Schwartzman, 1993). Indeed, diffusion modelling of microbial soil CO_2 has lead Keller and Wood (1993) to suggest that Proterozoic weathering rates may have been very similar to modern ones. This is strongly supported by a consideration of the likely trend of p_{CO_2} for pre-land-plant atmospheres; recent models indicate substantially higher p_{CO_2} and hence higher chemical weathering rates for these times (Berner, 1995).

2.9 Physical weathering

Rocks are slowly exhumed to the surface by erosion. The elastic rock volumes experience unloading, whereby deep conditions of very high, but balanced, triaxial stresses are replaced by *in situ* biaxial compressive stresses of up to 35 MPa, in excess of most rocks' strength at the surface. The existence of these stresses is vividly illustrated (see Folk & Patton, 1982) by geologically sudden unloading produced by quarrying or mining with resultant sudden sheeting, arching or catastrophic rock bursting (the greatest fear for mine geologists). Slower unloading brought about by surface erosion produces a variety of rock joints, other fractures and curvilinear exfoliations and sheeting joints parallel to buttressed rock outcrop or valley surfaces (Holzhausen, 1989). These various imperfections are then acted upon by chemical weathering solutions, by biogenic attack and by other physical processes, chief of which are freeze–thaw (frost) and salt weathering. Both of the latter involve the production of stresses through crystallization of solids from solution in tiny rock fractures or near-isolated pores. The net effect, in combination with gravity, frost shattering, salt growth and near-surface chemical weathering, is the production of talus and colluvium. In this way the landscape reduces, almost by autodestruction. Two examples of weathering are shown in Fig. 2.17.

Frost weathering is particularly important in high mountainous catchments not under permafrost. It is an effect only partly due to anomalous expansion and decreasing density as water freezes. The accompanying increase in volume of up to 13.5% (at $-22\ °C$) generates up to about 200 MPa of pressure in confined situations (Bridgman, 1911). However, the story is inevitably more complicated (Walder & Hallett, 1985; Tharp, 1987) because one has to consider the effect of ice whisker crystal growth, liquid film transfer and stress gradients in tiny to small isolated or tortuous cracks (< 0.5 mm) and pores. Additional stresses—as much as 10 times those arising from simple expansion above—may arise from ice growth as clusters of parallel ice crystal needles grow normal to the freezing surface. Provided the small crack is supplied with a net input of water, then the stresses arising from crystal growth are limited only by the tensile strength of water, which is drawing water molecules to the ends of the growing crystals through capillary films. This process carries on way below normal freezing point in the thin films, being most effective between -4 and $-15\ °C$.

An interesting long-term experimental research programme into frost shattering is summarized by Lautridou and Ozouf (1982). It seems that frost weathering is most effective in tiny cracks and crevices of irregular shape in temperate to subarctic climates where repeated thawing and freezing occur on a daily basis, the water moving around by capillary attraction. Between 0.02 and 0.8% by mass of shattered material was generated in experimental 40 yr shatter cycles with a variety of 'hard rocks' (Lautridou & Seppala, 1986). Porous and permeable rocks, where drainage is freer, are not so prone to shattering. Up to 165 freeze–thaw cycles per year have been recorded in the French Alps at altitudes of just 2.5 km. Anyone who has spent any time in high mountains with rocks whistling round their ears will attest to the very high fragmentation rates due to this mechanism. A certain amount, sometimes a significant amount, of daily freeze–thaw also occurs in low-latitude and/or high-plateau deserts where winter rains and dews may provide enough moisture for the freeze–thaw process to be effective.

Salt weathering (see the authoritative review by Cooke *et al.*, 1993) has been greatly underestimated as a weathering type. It occurs in semiarid to arid climates and in coastal areas of all latitudes where salts are concentrated and where dews, coastal mists, sea spray and ordinary rainfall provide the necessary liquid phases. There seem to be three ways in which salt expansion may give rise to stresses that lead to rock disintegration: (i) periodic hydration and dehydration, (ii) periodic heating and cooling, and (iii) crystal growth (Goudie *et al.*, 1970; Cooke, 1979;

(a)

(b)

Fig. 2.17 (a) Granodiorite corestones in the weathering zone, Tai Po, Hong Kong (b) This was once solid granite, now being tested by a hand-inserted shear vane. (Photos courtesy of S. Hencher.)

Goudie, 1985; Sperling & Cooke, 1985). The last mechanism is probably the most important. As we have seen, crystal growth stresses are particularly dependent upon rock porosity and the mechanism is most effective in porous sedimentary hosts. Salts vary in their ability to disintegrate rocks by crystal growth, sodium and magnesium sulphates being most effective. Crystal growth stresses occur in tortuous cracks under pressure. They arise at the crystal–rock interface via aqueous films as the crystallizing salts occlude the available space. Open systems, where salts crystallize due to evaporation, cannot give rise to changes of volume and hence cannot give stresses.

A neglected form of physical weathering occurs when sedimentary pyrite is oxidized in fissile shales. The resulting crystal growth of ferrous sulphate and gypsum is a highly efficient method of rock disaggregation.

The net effect of all the physical processes noted above is the production of spalled rock fragments and rock/mineral grain debris. It is now considered impossible for diurnal temperature changes in deserts to cause rock exfoliation since many experiments have failed to reproduce the effects in the laboratory. The effect is ascribed to the spalling-off of partly chemically weathered rock skins from bedrock surfaces, either beneath a soil layer (Fig. 2.17) or often concentrated on the shady parts of rock surfaces. Here, the effects of surface moisture upon minerals is greatest and acts for longer periods.

2.10 Soils as valves and filters for the natural landscape

In this chapter we have examined the weathering process in some detail, for this is how clastic sediment and dissolved ions originate. We have not yet considered soil specifically. It is useful to look at soil as a graduated carapace on landscape through which atmosphere and hydrosphere communicate with lithosphere. It is thus a valve in the sense that it allows some proportion of precipitation to pass through it and hence into the saprolite to further the weathering process. It is a filter in the sense that the (usually) well-drained, aerated and oxygenated surface soil layers are the site of a myriad of chemical reactions, many of them biologically mediated, that modify (filter) the intercepted precipitation. Biological mediation in this case includes not just the obvious respiratory capacity of higher-plant root systems and the ingestive capacity of worms, but the huge metabolizing soil microbe populations of bacteria, actinomycetes, mycorrhizal fungi (termed 'rock-eaters' by Jongmans *et al.*, 1997), algae, protozoa and slime moulds.

In view of the above comments it will come as no surprise to the reader that there are in fact very many possible soil types, for bedrock, climate, altitude and the biological filtering factors just mentioned all play their roles in determining the exact nature of the soil carapace. But it is not necessary for sedimentologists to get bogged down in the intricacies of soil classifications. Soil was mentioned previously as graduated, meaning it commonly divides up into more-or-less distinct layers. It is the nature of these layers that forms the basis for soil classifications. Generalizing again, there is often an upper accumulation of organic litter of dark hue. Then there is a middle horizon that is depletive as a result of mineral alteration, weathering and physical washing-out of materials (ellutriation). Finally there is a lower accumulative horizon that represents the reprecipitation of dissolved materials from above or the site of deposition of clay-grade material from throughflowing waters (illuviation). In older classifications these three layers are given the codes A, B and C respectively.

An important point is that natural soils must develop progressively with time, leading to changes in chemistry and physical characteristics. Residual soils, particularly duricrusts, occur in landscapes from which only small amounts of soil breakdown products are removed physically. Such soils are extremely useful to geologists in assessing the nature of past climates and of climate change. They are also, importantly, indicative of rather stable tectonic regimes, for the occurrence of remnant duricrusts on ancient geomorphic surfaces is an important indication of

pretectonic uplift, climate and weathering (see Valeton, 1996). In eroding landscapes the extent and age of the soil will depend on the balance between formation and erosion. Studies in landscapes like those of Scandinavia, whose soils were initiated on ground-up mineral grains after 'ice-out' as the Quaternary glaciers retreated, enable estimates of rates of elemental depletion and soil textural development with time (Sverdrup & Warfvinge, 1995).

Following Chesworth (1992) we may grossly simplify chemical weathering and soil environments into acid, alkaline and reduced trends (Figs 2.18 & 2.19). These occur in humid and arid climates and in wetland systems, respectively. We may also simplify soil types into four major groups, ignoring the distinctive but ephemeral character of soil developed upon unconsolidated, glass-rich volcanic ashes. A further subdivision is useful in the field description of palaeosols in the sedimentary rock record (Fig. 2.20; Mack *et al.*, 1993), but here it is important to be able to assess the degree of secondary alteration during burial of what were once primary soil features. Thus physical attributes such as soil nodules and agglomerations (peds) and expansion/contraction cracks are easily destroyed (unless mineralized) by compaction during burial. For a mineralogical example, consider the palaeosol profile of Carboniferous age investigated by Wahlstrom (1948) in a classic study. Here the common occurrence of illite and K-feldspar in the upper part of the profile is clearly anomalous, suggesting to Nesbitt and Young (1989) that pervasive K addition has occurred during burial.

Acid weathering trends occur with positive water budgets, causing hydrolysis and leaching, and resulting in ferralitic and podzolic soils:

1 Ferralitic weathering and soil formation today are characteristic of stable well-drained landscapes in humid tropical climates. Here, pH conditions grade from moderately alkaline to mildly acid, the former conditions encouraging silica solubility. Mature soils ($> 10^4$ yr) are dominated by aluminium (bauxites) and iron (laterites) oxides and hydroxides, with the minerals goethite, haematite, kaolinite and gibbsite.

2 Podzolic weathering and soil formation occur in cool, humid temperate climates and widely in high mountains. Humus accumulates rapidly in an upper horizon since under cool conditions the rate of metabolic breakdown is slow. Moderately to highly acidic

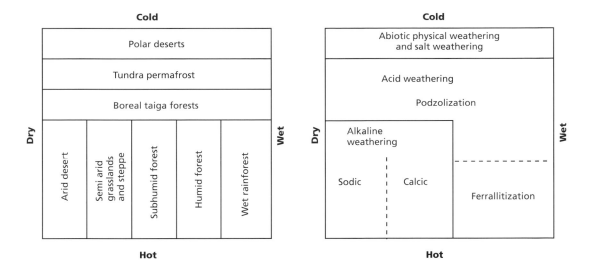

Fig. 2.18 Simple box classification of climate, vegetation, weathering and soil types developed on well-drained (i.e. nonreducing) substrates. This scheme ignores local (sometimes crucial) control by particular lithologies. (Simplified after Chesworth, 1992, and sources therein.)

conditions lead to Al and Fe leaching in middle, pale Si-rich layers and their reprecipitation in lower layers.

Alkali weathering trends occur in arid, semiarid and Mediterranean areas with annual water deficit, and include calcic and sodic soil types:

1 Sodic soils occur in arid climates where there is an upward movement of porewaters leading to evaporative concentration of alkali and alkaline-earth elements as sulphates, bicarbonates and chlorides. They often occur in interior drainage catchments and reflect ions translocated to the soil profile from considerable distances in throughflowing groundwaters. They are a major problem in irrigated farmlands with poorly planned drainage.

2 Calcic soils occur widely in semiarid and Mediterranean climates where they result from a balance between restricted downward leaching of $CaCO_3$ (often derived from aeolian dust) and its reprecipitation

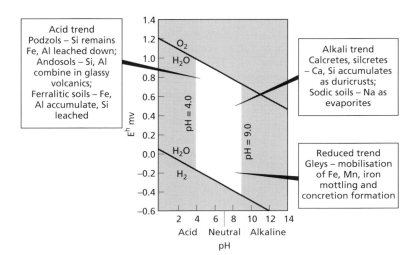

Fig. 2.19 Broad classification of soil chemical processes in terms of E^h and pH. (After Chesworth, 1992.)

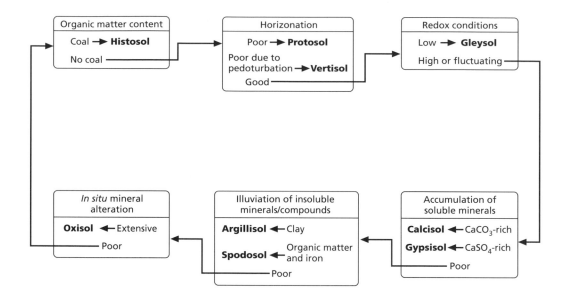

Fig. 2.20 Simplified flowchart of palaeosol orders. (From Mack et al., 1993.)

as calcite in lower soil layers. High pH leads to aggressive dissolution of silicate minerals in parent materials. Stable isotope analysis (C, O) of calcic soils gives valuable evidence for palaeoclimate regime (see Mack *et al.*, 1994; Andrews *et al.*, 1998).

Further reading

My favourite standard chemistry text, with material relevant to some of the chemical principles used in the study of silicate weathering, is by Atkins (1992). Readers lacking college-level chemistry and wanting texts with real geological and environmental relevance would do well to consult Gill (1989) and, especially, Andrews *et al.* (1996). Stumm's (1992) fundamental book on chemical weathering is made of sterner stuff. The collection of papers on the chemistry of rock weathering edited by White and Brantley (1995) is invaluable, and a bargain at the price. Selby (1992) is very good on physical weathering from a landscape perspective.

3 The inorganic and organic precipitation of sediment: chemical, biochemical and biological

. . . changing school, sandstone changed for chalk
And ammonites for the flinty husks of sponges,
Another lingo to talk
And jerseys in other colours.

Louis MacNeice, 'Autumn Journal', *Collected Poems*, Faber

3.1 Marine and freshwater chemical composition: chemical fluxes to and from the oceans

The oceans are vast sinks for dissolved chemical and biochemical weathering products. Early estimates of the age of the Earth assumed that the oceans behaved passively, receiving but not getting rid of elements like sodium. Thus in 1899 the Irishman Joly calculated on this basis that the Earth was 9×10^7 yr old. The present ocean is now thought to be in a state of dynamic equilibrium, with the continuity equation satisfied, as everywhere, by the balance of inputs from weathering and mid-ocean-ridge activities and outputs of chemical and biological precipitation. However, it is a major research issue as to whether there have been secular variations in ocean water chemistry over periods of 10^7–10^8 Myr in the geological past (see Hardie, 1996).

The oceans are well mixed chemically, with mean residence times calculated to be of the order of 10^3–10^4 yr or so (see Fig. 1.1). This broad chemical homogeneity is accompanied in detail by a number of local contrasting physical gradients of temperature and density arising from so-called conveyor belts of heat energy from poles to equator and vice versa. These cause some disparities between the residence intervals in separate water masses, perhaps by factors up to two orders of magnitude.

Table 3.1 reveals that:
1 Seawater contains about 300 times more dissolved solids than does fresh water.

Table 3.1 Comparison of the compositions of 'average' river-water and 'average' seawater. Note that in seawater, with its high ionic strength, not all of any given anion or cation is available for reaction because of the phenomena of complexing and ion pairing. The flux of chemical species from the world's rivers into the oceans may be calculated by making use of the measured concentration of elements in river-water and the mean annual discharge of the rivers involved. To this we must correct for the flux of oceanic wind-blown salts to the continents and for the effect of various pollutants. Chief amongst the latter is SO_4^-, whose proportion provided by pollutants is estimated to be 20–30%. (Data collected from Livingstone, 1963; Drever *et al.*, 1988.)

Ion	Seawater conc. (mol/kg)	Order of abundance	River-water conc. (mol/kg)	Order of abundance	Net river flux—no cyclic salts or pollution (10^{12} mol/yr)	Sea/river
Na^+	0.47	2	2.7×10^{-4}	4	5.91	1740
K^+	1×10^{-2}	5	5.9×10^{-5}	7	1.17	170
Ca^{2+}	1×10^{-2}	5	3.8×10^{-4}	2	12.36	26
Mg^{2+}	5.4×10^{-2}	3	1.7×10^{-4}	3	4.85	318
Cl^-	0.55	1	2.2×10^{-4}	4	3.27	2500
SO_4^{2-}	3.8×10^{-2}	4	1.2×10^{-4}	6	3.07	317
HCO_3^-	1.8×10^{-2}	6	9.6×10^{-4}	1	32.09	1.9
pH	7.9		~7			
Ionic strength	0.65		0.002			

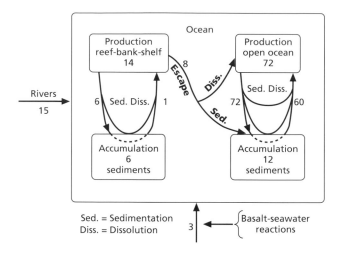

Fig. 3.1 Global calcium carbonate cycle. (From Mackenzie & Morse, 1992.) Carbon fluxes are in units of 10^{12} mol/yr as $(Ca,Mg)CO_3$.

2 Seawater cations and anions in decreasing order of abundance are Na^+, Mg^{2+}, Ca^{2+}, K^+ and Cl^-, SO_4^-, HCO_3^-. Freshwater cations and anions in decreasing order of abundance are Ca^{2+}, Na^+, Mg^{2+}, and HCO_3^-, SO_4^{2-}, Cl^-.

3 River fluxes are dominated by Ca^{2+} and HCO_3^-, both of which take part in active recycling via carbonate reactions.

Thus seawater is not simply 'concentrated' riverwater, because of the different proportions of solutes. Observed seawater composition results from:

• inputs reflecting composition of continental runoff;
• chemical and biological fractionation by mineral reactions taking place in the water itself, and around the sediment–water interface;
• burial of oceanic waters as 'porewaters' in sediment;
• seawater–basalt reactions along the ocean ridges.

Secular seawater compositional changes can come about if the balance of input–output is changed. Calculation of the magnitude of the various fluxes into and out of seawater leaves significant anomalies, which reflect the importance of hydrothermal basalt–seawater reactions at the world's ocean ridges. Pluto as well as Neptune plays a role in the oceanic balance sheet!

3.2 The carbonate system in the oceans

First of all, consider the global carbonate cycle (Fig. 3.1). Each year some 15×10^{12} mol of carbon are brought to the oceans as bicarbonate. Annual production of carbonate from the oceans is far greater,

at about 86×10^{12} mol. Some of the difference is made up by calcium provided at the mid-ocean ridges, but the greatest discrepancy is explained by the oceanic uptake of carbonate derived from deep-water dissolution processes. Thus the annual deposition of calcium carbonate from the oceans, as chemical and (mostly) biologically mediated precipitate and skeletons, is only about 18×10^{12} mol, with about one-third of this sourced in the shallow-water 'carbonate factory' and the majority from the oceanic surface-water 'factory'.

Now consider the chemical aspects of the carbonate cycle, for carbonate reactions play a key role in maintaining seawater pH. Seawater is well buffered and has a pH in the narrow range of 7.8–8.3. Any tendency to increase surface seawater acidity, caused for example by higher atmospheric CO_2, is opposed by the following reactions: first, the slow dissolution reaction of CO_2 with water to give the weak carbonic acid,

$$H_2O + CO_2 \rightleftharpoons H_2CO_3$$
$$H_2CO_3 + CaCO_3 \rightleftharpoons Ca^{2+} + 2HCO_3^- \quad (3.1)$$

and secondly, the rapid ionization of carbonic acid in two stages, the first more important than the second,

$$H_2CO_3 \rightleftharpoons H^+ + HCO_3^-$$
$$HCO_3^- \rightleftharpoons H^+ + CO_3^{2-} \quad (3.2)$$

These two reactions balance the excess positive cation charge in seawater (neglecting the tiny contribution made by the H^+ ions). At normal oceanic pH of

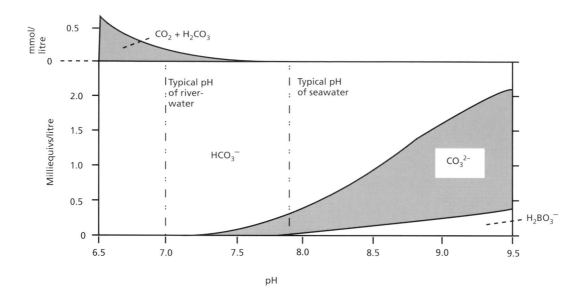

Fig. 3.2 Variation of alkalinity components of seawater with pH. (After Cloud, 1962.) The 'typical' values for seawater and river-water (approximate) have been added.

Table 3.2 Solubility products (K) and ion activity products (IAP) for calcite, dolomite and aragonite in seawater at 25 °C. Solubility products for calcite and aragonite after Berner (1971), for dolomite after Hsu (1966). IAP for $CaCO_3$ is an approximate mean for surface seawater at 25 °C (Berner, 1971). IAP for dolomite after Hsu (1966). Note that the surface seawater is supersaturated with respect to all the chief carbonate minerals.

Mineral	K	IAP	IAP/K (= Ω)
Calcite	4.39×10^{-9}	1.35×10^{-8}	3.4
Aragonite	6.3×10^{-9}	1.35×10^{-8}	2.1
Dolomite	$\sim 1.0 \times 10^{-17}$	1.0×10^{-15}	~ 100

around 8, more than 80% of dissolved carbon is in the form of HCO_3^-, with CO_3^{2-} making up the remainder (Fig. 3.2). Not all of the bicarbonate is available for reaction because of the involvement of the anion with Mg^{2+} in the ion pair $MgHCO_3^+$. Any increase in alkalinity of seawater is opposed by the buffering reaction:

$$HCO_3^- + OH^- \rightleftharpoons H_2O + CO_3^{2-} \qquad (3.3)$$

Let us briefly consider the role of ionic strength. The solubility of calcium carbonate in pure water may be calculated from thermodynamic data. It is found, however, that the solubility is very much greater in a solution like seawater where other abundant ionic species are present. These ionic species together with polar H_2O molecules tend to cluster around the oppositely charged ions of Ca^{2+} and CO_3^- preventing the ions coming together to precipitate, i.e. the solubility increases. The greater the charge on the 'pollutant' ions, the greater the effect. Thus solutions like seawater are said to have a high ionic strength due to the formation of complex ions and ion pairs. Seawater has an ionic strength of about 0.7 whereas fresh water

has values around 0.002. Because of this difference in ionic strengths, the solubility of $CaCO_3$ in seawater is much higher than in fresh water.

Surface seawater is distinctly supersaturated with respect to aragonite, calcite and dolomite (Table 3.2), yet precipitation of inorganic $CaCO_3$ is confined to subtropical and tropical locations where it is probably less important worldwide compared to biogenic fixation of $CaCO_3$, chiefly by shallow-water coral reefs and open-ocean planktonic algae. The level of carbonate supersaturation decreases with increasing latitude and decreasing surface seawater temperatures (see Fig. 3.16; Burton & Walter, 1987; Opdyke & Wilkinson, 1993).

Calcium carbonate exists as the mineral polymorphs calcite and aragonite. Both may form as inorganic precipitates or as biological secretions in the hard parts of numerous organisms. Aragonite does not usually precipitate from fresh water. Properties of the two minerals are listed in Table 3.2. Aragonite is metastable under earth-surface conditions, being a high-pressure equilibrium carbonate, as found in blue schist metamorphic facies. Many of the problems of carbonate sedimentology revolve around the timing and chemical constraints upon aragonite vs. calcite precipitation, aragonite inversion to calcite, and aragonite dissolution. An important consequence of the charge similarity and ionic radius of Ca^{2+} and Mg^{2+} ions and of the structure of the calcite lattice is that Mg^{2+} may substitute extensively for Ca^{2+} in calcite. The calcite formula is therefore more correctly written as $(Ca_{1-x},Mg_x)CO_3$ where x is commonly in the range 0.01–0.25, never exceeding about 0.4. These calcites with more than 5% $MgCO_3$ are known as high-magnesian calcites; they are the most reactive and most soluble of all the various sedimentary carbonates. The Mg content of certain organic hard parts seems to be related to temperature, with greater amounts of Mg in warmer waters. Small amounts of Fe^{2+} (up to a few thousand parts per million) may also substitute for Ca^{2+}, giving rise to ferran calcites in low-E_h conditions. The substitution of trace amounts of Mn^{4+} causes calcite to luminesce under the influence of cathode-ray bombardment. If the flux of Mn^{4+} varied with time during calcite crystallization, then luminescence studies reveal tell-tale growth zones, which may often be mapped out in stratigraphic sections.

The aragonite lattice cannot take up Mg^{2+} when it is an inorganically precipitated phase, although certain skeletal aragonites in corals may contain up to about 0.001% Mg^{2+}. Strontium may be taken up in the aragonite lattice to a maximum concentration of about 10 000 ppm. The crystal habit of aragonite in precipitated and some biogenic phases (e.g. calcareous algae) is usually of fibrous type, whilst calcite may be fibrous, rhombic or the familiar 'dog's tooth' morphology.

The double carbonate dolomite, $CaMg(CO_3)_2$, is mainly a diagenetic mineral. Ferrous iron may substitute for Mg^{2+} in the solid-solution series dolomite–ankerite $(CaMg_{0.75}Fe_{0.25})(CO_3)_2$. The pure calcium–iron carbonate is unknown in Nature. The dolomite lattice is highly ordered, with alternating layers of cations and CO_3^{2-} groups, in which the cation layers are alternatively Ca^{2+} and Mg^{2+}. Dolomites fall into two broad groups: generally finely crystalline calcian dolomites (> 53% Ca) and near-stoichiometric fine to coarsely crystalline dolomites.

Recent shallow-water tropical and subtropical calcium carbonate deposits are predominantly composed of aragonite and high-magnesian calcite, whilst temperate shallow carbonates contain dominantly calcite. Recognition of carbonate minerals is greatly aided by staining techniques in thin section and peels. Calcite, Fe-calcite, high-magnesian calcite, aragonite, dolomite and Fe-dolomite may all be distinguished in this way.

Further information on basic carbonate mineralogy

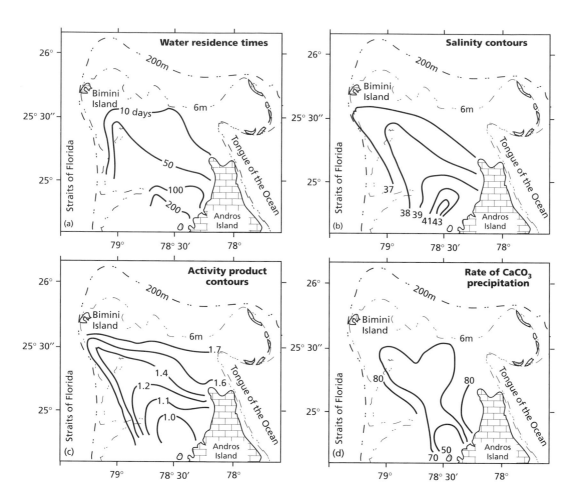

Fig. 3.3 Bathymetric maps. (All after Broecker & Takahashi, 1966.) (a) Mean summer residence time of water on the Bahamas Bank. (b) Typical values of salinity (parts per thousand). (c) Activity product contours for $[Ca^{2+}][CO_3^{2-}]$. (d) Calculated rates of $CaCO_3$ precipitation (mg/cm² yr).

Calcium carbonate deposits, held by some authors to be of inorganically precipitated origin, occur in several areas, notably the Bahamas, Arabian Gulf and Dead Sea. In each case the mineral form is aragonite in the characteristic needle-like crystal habit with crystal sizes of a few micrometres. In the Dead Sea, events of mass precipitation have been correlated with the appearance of the ghostly 'whitings': large patches of aragonite suspensions that appear suddenly in the surface waters. The Dead Sea whitings (Neev & Emery, 1967) are characterized by an immediate decrease in HCO_3^- in the water mass, indicating

$CaCO_3$ precipitation by the reverse of reactions in eqn (3.1). Chemical data are not available for the Arabian Gulf occurrences, but the size and nature of the whitings leave little room for an alternative explanation to inorganic precipitation. Increase in CO_2 uptake during periodic diatom 'blooms' has been advanced as the cause for these whitings. A small problem here, though, is the rarity of preserved aragonite in the bottom sediments of the offshore Gulf.

Classic studies on $CaCO_3$ precipitation in the waters of the Great Bahamas Banks were undertaken by Cloud (1962) and Broecker and Takahashi (1966; Fig. 3.3). The physical chemistry and water dynamics in these saturated shallow subtropical seas were refined by Morse *et al.* (1984) (but abandoned after disruption by armed drug smugglers). These authors confirm the main conclusion of the early studies, that a significant percentage of the calcium carbonate

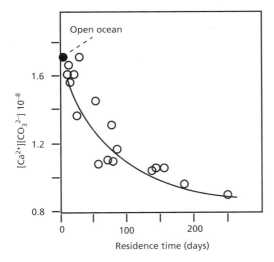

Fig. 3.4 The relationship between activity product and water residence time for the Bahamas Banks. (After Broecker & Takahashi, 1966.)

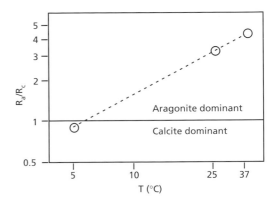

Fig. 3.5 Relative precipitation rates of aragonite vs. calcite (R_a/R_c) as a function of temperature in seawater five times supersaturated with respect to calcite. (After Burton & Walter, 1987.)

on the banks is inorganically precipitated. More recent studies of trace-element compositions, chiefly Sr (Milliman *et al.*, 1993), and careful field observations of whiting plumes (Shinn *et al.*, 1989) support these chemical studies. As cool Atlantic water passes onto the shallow Bahamas Banks, it is warmed up, salinity increases and $CaCO_3$ decreases, indicative of precipitation.

The rate of loss of $CaCO_3$ is obtained by dividing the $CaCO_3$ deficit by the mean residence time on the Banks for a particular water sample (Fig. 3.4). Mean residence was originally calculated (Broecker & Takahashi, 1966) knowing the degree of incorporation of atomic-bomb-produced ^{14}C from the atmosphere. Mean precipitation rates of $CaCO_3$ thus obtained are around 50 mg/cm^2 yr. These gross chemical trends indicate an important degree of inorganic contribution of $CaCO_3$ in the area, perhaps as much as 50%.

3.3 Advances in understanding carbonate reaction kinetics and their significance

Even though seawater is supersaturated with respect to aragonite, calcite and dolomite, only the former mineral seems to precipitate directly from seawater (excluding porewaters from the present discussion), and that with some difficulty (as we have seen above,

particularly when $\Omega < 4$). Why should this be so? First of all we should note that there is a very strong temperature control on polymorph selection. For any given ratio Mg/Ca > 1, aragonite is always favoured by higher seawater temperature (Fig. 3.5; Burton & Walter, 1987; Morse *et al.*, 1997).

Secondly, numerous experimental and theoretical studies indicate that, at low supersaturations, calcite and aragonite precipitation is inhibited by other dissolved ionic species. Chief amongst these carbonate 'poisons' are Mg^{2+} (Pytkowicz, 1965; Berner, 1975; Berner *et al.*, 1978), Fe^{2+} (Meyer, 1984; Sumner & Grotzinger, 1996) and dissolved phosphate (Simkiss, 1964; Griffin & Jurinak, 1973; Morse, 1974; Reddy, 1977; Berner *et al.*, 1978; Burton & Walter, 1990; Dove & Hochella, 1993). Concerning p_{CO_2} and Mg^{2+} ions:

1 Changing p_{CO_2} (for fixed degrees of supersaturation) has little effect upon the rate of calcite or aragonite precipitation.

2 Dissolved Mg^{2+} in seawater has no effect upon the seeded precipitation of aragonite (Fig. 3.6).

3 Dissolved Mg^{2+} severely retards calcite precipitation (Fig. 3.6).

4 Calcite precipitated from seawater on *pure* calcite seeds contains 7–10% $MgCO_3$ as an overgrowth of the more soluble high-magnesian calcite (see also Wollast *et al.*, 1980).

5 In Mg^{2+}-deficient 'seawater', with < 5% of the normal Mg content, Mg^{2+} does not appreciably retard the seeded precipitation of calcite. Low-magnesian calcite is thus stable in freshwater regimes of suitable pH.

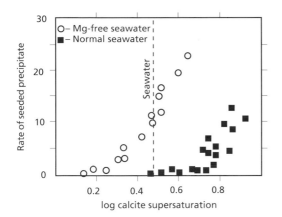

Fig. 3.6 The effects of magnesium on calcite precipitation. (After Berner, 1975.) No influence of magnesium was seen in similar experiments involving aragonite precipitation.

The surface 'poison' effect of Mg^{2+} upon calcite precipitation may be due to the fact that the smaller Mg^{2+} ions are more firmly hydrated by polar water molecules than are Ca^{2+} ions. More thermodynamic work must be done to dehydrate these Mg^{2+} ions than to dehydrate Ca^{2+} for the growth of aragonite lattices (Bathurst, 1968; Lippman, 1973). Mg^{2+} has no effect upon the formation of the orthorhombic aragonite lattice.

Both aragonite and calcite precipitation are inhibited by the trace quantities of dissolved phosphate ions present in most seawater. It has been shown (Burton & Walter, 1990) that calcite and aragonite precipitation rates depend upon both the saturation state of calcium carbonate and the phosphate concentration. With no phosphate around, aragonite precipitates much faster than calcite. In normal phosphate-bearing seawater, the relative rates depend upon pH through the balance between PO_3^{3-} and HPO_4^{2-}. At normal seawater pH of around 8, the most stable phase to precipitate is calcite, but we have already seen above that this phase is actively discouraged by Mg^{2+} poisoning. The pH control is probably more important in sediment porewaters where pH is more variable.

Novel experimental techniques in the past few years, particularly the advent of atomic force microscopy (AFM) and scanning force microscopy (SFM) imaging, have shed light on the nature of crystal growth for the comparatively simple calcite lattice. It appears (Gratz *et al.*, 1993) that growth occurs

as advancing monomolecular steps on the calcite cleavage surfaces prepared for the experiments. Step nucleation occurs mostly at growth spirals, and the steps are fed by addition of material from the growth solution, not by surface diffusion as observed in the vapour growth of many artificial monatomic semiconductors and metals. SFM experiments (Dove & Hochella, 1993) on freshly prepared calcite growth surfaces, as distinct from cleavage surfaces, at saturations ($\Omega > 1–2$) appropriate for seawater reveal that precipitation begins with scattered surface nuclei. These nuclei are initially high and grow, spread and combine to form eventually beautifully layered growth patterns resembling spirals whose heights are only a few monolayers (Fig. 3.7). Growth rates decay from initial to layered stages, the longer-term equilibrium rates being linear (first order). A very important observation during the experiments was that exposure of growing crystals to air, even for short periods of time, caused re-establishment of rapid spot-nucleated growth. Thus those natural carbonates that form under conditions of frequent wetting–drying cycles (e.g. calcisols, beachrocks, etc.) experience rapid growth and the production of a highly porous fabric and low density. Rate laws for these environments will be of higher order than for the simple linear kinetics of a continuously wetted environment.

The AFM and SFM experiments shed light on the effects of phosphates upon precipitation. Addition of phosphate during the nucleation stage changes the nuclei to amorphous shapes, the phosphate perhaps blocking growth of the $CaCO_3$ nuclei. The addition of phosphate during the layer stage of precipitation disrupts the growth of regular spiral steps and terraces, perhaps because the phosphate ions block spiral growth sites like kinks, steps or holes (Fig. 3.7; Dove & Hochella, 1993; Gratz *et al.*, 1993).

The lack of dolomite precipitation in modern seawater must also result from an inhibition mechanism, since from the known composition of seawater dolomite should be the first mineral to precipitate (Hsu, 1966). The very high degree of ordering in the dolomite lattice seems to result in extremely slow crystal nucleation and growth rates. Attempts at laboratory precipitation of dolomite result in the formation of more poorly ordered and metastable magnesian calcites of dolomite composition, known as protodolomites. A surface poisoning effect, when hydrated Mg^{2+} ions surround the Ca^{2+} growth planes

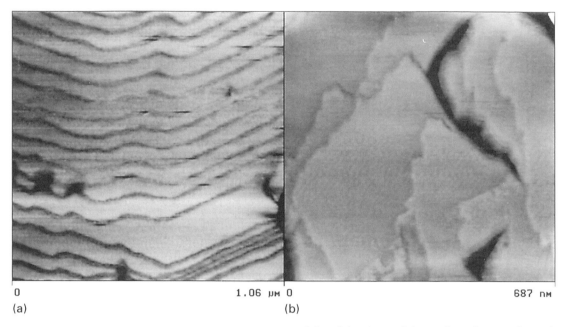

0   1.06 µm 0   687 nm
(a)   (b)

Fig. 3.7 SFM photos showing: (a) calcite growth in smooth spiral steps, each 3 Å step being individual monolayers of sheets of carbonate and calcium groups on a growth surface; and (b) the effect of adding phosphate in dilute solution to a similar stepped surface to that in (a)—the large, jagged, irregular steps coincide with inhibited and irregular growth. (From Dove & Hochella, 1993. Photos courtesy of P. Dove.)

Fig. 3.8 Secondary electron images showing (a) nanobacterial ferroan dolomite coatings and (b) coatings on possible twinned nanobacteria caught in the act of cell division. (From Vasconcelos & McKenzie, 1997. Photos courtesy of J. McKenzie.)

of tiny dolomite nuclei, may impede growth much as postulated for the inhibition of calcite growth above. Recent studies indicate that significant dolomite precipitation is microbially mediated and occurs under shallow subsurface anoxic conditions (Mazzullo *et al.*, 1995; Vasconcelos *et al.*, 1995; Vasconcelos & McKenzie, 1997; Fig. 3.8).

Despite the evidence for lack of primary dolomite precipitation from the modern oceans, it has long been suspected that dolomite was precipitated from Precambrian seas. Not only are dolomitic rocks much commoner in the Precambrian, but careful petrographic and stable isotopic studies (e.g. Tucker, 1982)

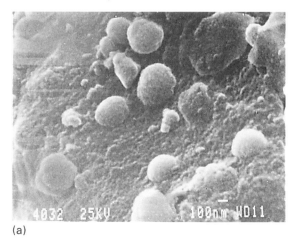

(a)

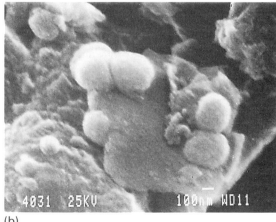

(b)

Fig. 3.9 Characteristic fan-shaped seafloor growths of primary aragonite (now pseudomorphed by calcite). Scale bar = 5 cm. From > 2 Ga, Reivilo Fn, South Africa. (From Sumner & Grotzinger, 1996. Photo courtesy of D. Sumner.)

suggest that the pristine condition of dolomitic ooids, pisoliths and various cements could not have resulted from any diagenetic recrystallization. It is likely that there have been major secular changes in seawater chemistry through geological time (see Hardie, 1996, for a recent persuasive account). For example, Archaean and Proterozoic oceans are thought to have been deficient in O_2, with abundant Fe^{2+} acting as a 'poison' to calcite precipitation. Under such conditions, Sumner and Grotzinger (1996) suggest that widespread seafloor precipitation of aragonite occurred, with characteristic morphologies (Fig. 3.9). Data from the composition of marine authigenic carbonate minerals (see Wilkinson *et al.*, 1985; Fig. 3.10) and evaporites indicate that the oceans may have oscillated between two states, aragonite oceans and calcite oceans.

Aragonite ocean states (most notably in the Pennsylvanian/Permian) corresponded to 'icehouse Earth' conditions as today, with widespread precipitation of aragonite rather than calcite in the low-latitude oceanic waters. Conditions are marked by high Mg/Ca ratios and the dominance of magnesium sulphate potash evaporite minerals (see further below). Calcite oceans dominated in Lower Palaeozoic and Mesozoic 'greenhouse Earth' times and were characterized by low Mg/Ca ratios, calcite precipitation and potassium chloride potash evaporite minerals.

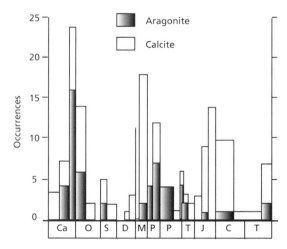

Fig. 3.10 Variation in calcium carbonate mineralogy with time for chemically precipitated (authigenic) minerals. (From Wilkinson *et al.*, 1985.)

3.4 Pre-Recent and future CaCO₃ reactions

It is unlikely that oceanic pH has ever strayed far outside the range 6–9. Thus calcite is present in rocks as old as 3.8 Gyr, and a pH of less than 6 is thus improbable (other Ca-bearing phases would have precipitated, for which there is no evidence) unless the p_{CO_2}

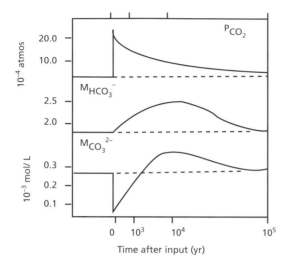

Fig. 3.11 Thought experiment involving projected time changes in the distribution of carbonate species in the atmosphere and oceans as a result of the instantaneous burning of all fossil fuels and the dissolution of the resulting CO_2 in the oceans. (From Walker & Drever, 1988.)

value was unreasonably high (> 1 atm). Oceanic pH > 9 is unlikely because massive sodium carbonate precipitation (see below) would have occurred, for which there is also no evidence.

The Mg poisoning effect upon modern inorganic calcite precipitation led Sandberg (1975) to attribute Palaeozoic calcitic ooids with perfect preservation originally to precipitation from seawater low in Mg. However, it was subsequently suggested (Pigott & Mackenzie, 1979; Sandberg, 1983) that calcitic ooids formed due to periods of higher p_{CO_2}. An increase of atmospheric p_{CO_2} above the present world average of around 320 ppm can be accommodated by increased levels of dissolved CO_2 (Fig. 3.11). Carbonate dissolves to compensate for the increased CO_2. This will initially favour the dissolution of high-magnesian calcites, which are significantly more soluble than pure calcite (Wollast *et al.*, 1980). Figure 3.11 presents the results of a thought experiment involving the instantaneous dissolution of all CO_2 produced by burning the entire fossil-fuel inventory of the planet.

Decrease of CO_2 would cause carbonate precipitation. The p_{CO_2} of modern surface ocean waters is controlled in large part by photosynthetic activities of planktonic plants (Peng & Broecker, 1991). The reduction in CO_2 content by such uptake depends

upon the efficiency with which nutrients like phosphates and nitrates are made available. The efficiency is high in most tropical and temperate seas, but low in the Antarctic and eastern Mediterranean. This is because of the lack of soluble iron, another essential cell-forming element, in these areas, the former due to lack of adjacent continental weathering, the latter to scavenging by sinking aeolian dust particles. It has been proposed, ingeniously, that the addition of iron to both oceans would help to mop up the excess CO_2 provided by future emissions (Martin *et al.*, 1990); but dynamic calculations do not support the idea as significant (Peng & Broecker, 1991).

3.5 Ooids

Ooids are one of a number of kinds of coated grain, their inclusion in this chapter as biochemically mediated sediment grains being somewhat controversial (see also Tucker & Wright, 1990). They are spherical to slightly ovoid, well-rounded carbonate particles (Figs 3.12 & 3.13) possessing a detrital inside (nucleus) and a concentrically laminated outside (cortex) of fine-grained aragonite and/or high-magnesian calcite often with many thin internal organic layers of algal origin (Shearman *et al.*, 1970). Diameters range from 0.1 to 2 mm or so. When the cortex is very thin, comprising one or two thin aragonite laminae, the ooid is said to be superficial. Usually individual aragonite laminae are up to about 10 mm thick. Ooids occur in marine sandwave and dune complexes in areas of strong tidal currents or in marine or lacustrine littoral or shallow sublittoral beach deposits. Radiocarbon studies reveal that modern ooids have grown very slowly, with surface layers giving ages only as recent as 225 yr, and with cores as old as 2300 yr.

Many sections through unaltered aragonite ooids (Fig. 3.12) show a pseudo-uniaxial cross under crossed polars, which gives a negative figure. This means that the predominant alignment of aragonite crystals must be with their *c*-axes tangential to the ooid surface. This is confirmed by X-ray and electron microscope analysis, the latter technique revealing that the aragonite crystals are in the form of rod-like particles 1–2 mm long with flattened end terminations (e.g. Loreau & Purser, 1973). In the Arabian Gulf, ooids from high-energy environments have a well-developed concentric arrangement of individual aragonite rods, whereas those from more sheltered areas have a radial

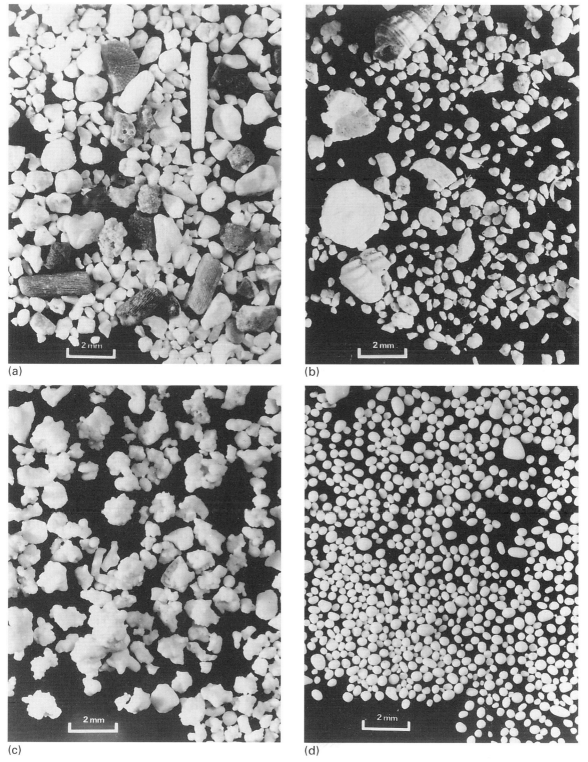

Fig. 3.12 Various carbonate grain types. (a) Skeletal carbonate grains, comprising abraded molluscan, foraminiferal, echinoid and algal fragments. (b) Skeletal–pelletal carbonate sand; note ovoid pellets, abraded foraminifera and gastropods. (c) Grapestone sand with aggregates composed of micritized skeletal grains and pellets. (d) Oolitic facies; note good sorting and high polish on individual ooids.

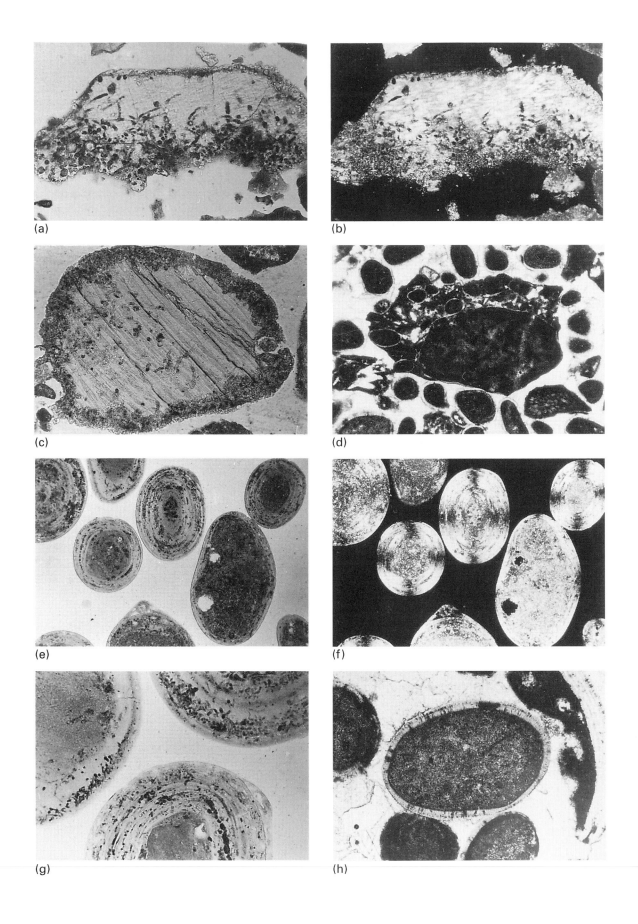

orientation with looser packing. Some ooids, notably those from the Great Salt Lake (USA) (Kahle, 1974; Sandberg, 1975; Halley, 1977; Medwedeff & Wilkinson, 1983), exhibit spectacular mixtures of radial aragonite, tangential aragonite and unorientated fabrics rich in clay minerals.

Regarding the vexed question of ooid genesis, Sorby's original snowball analogue stated that inorganically precipitated aragonite needles simply stuck on to the rolling exterior of an ooid. The lack of evidence for a sticky cohesive matrix on the smooth polished ooid outer surfaces has made this 'snowball' mechanism difficult to accept. We may be sure that high-energy environments somehow encourage tangential aragonite growth whilst low-energy environments encourage radial growth. Successful laboratory precipitation of alternating oolitic carbonate and organic laminae around spherules may occur where the concentration of organic material as humates is high and where organic membranes may form around suitable nuclei (Suess & Futterer, 1972; Ferguson *et al.*, 1978; Davies *et al.*, 1978). Experimentally produced carbonate laminae show radial aragonite growth under 'quiet' laboratory conditions and tangential fabrics after agitation. It is possible that aragonite crystals seed out from the ooid organic matrix to grow radially outwards, as in the growth of any uninterrupted crystal fabric from a solid surface. Such radial fabrics would be modified by turbulence and periodic abrasion into tangential fabrics in higher-energy environments.

The source of ooid aragonite needles remains a great problem since it is difficult to imagine tiny delicate crystals occurring in the bedload layer of high-energy water bodies such as tidal flows over oolite shoals. Paradoxical though it may be, the very process of intense bedload transport may encourage precipitation by a process akin to collision breeding of initial aragonite protocrystals. In suggesting this idea the author remembers observing collision breeding of

a mass of crystals from saturated salt solutions during experiments with a rotating Bagnold drum. Another point to consider might be the recognition that the ooliths spend much of their time stranded in the inside of moving bedforms such as sandwaves, dunes and ripples. Organic coats may develop here, causing radial aragonite laminae to begin to seed out from the porewaters between the oolite grains. Re-emergence of the grains into a turbulent bedload layer would then cause mechanical alignment of the aragonite rods into a tangential arrangement or would encourage tangential growth. Ooliths in quiet waters would remain with a radial structure. Periodic entrainment and burial would also explain the development of successive concentric layers.

By way of contrast to the above approach we may note Deelman's (1978) successful formation of Bahamian-type tangentially orientated ooliths from laboratory experiments with bicarbonate solutions. Intermittent stirring caused aragonite needles to precipitate around nuclei. Further agitation caused a high polish of the ooid surfaces and caused the aragonite needles to become well rounded at their edges. The tangential aragonite crystals were held together by surface bonding forces. These important results match well with the observed characters of natural ooids and of their physical environment of formation. The mucilaginous laminae present in natural ooids were not observed in the experimental examples, the solutions being simple bicarbonate ones. Doubtless these must form as the ooids undergo periods of 'rest' and become infested with blue-green algal slime.

The occurrence of calcitic ooids in the geological record with well-preserved radial microfabrics has posed problems of interpretation. Despite Sorby's acute original deduction that these were primary, the common Bahamian aragonite forms were considered to be the prototypes for most ancient ooids, an example of overzealous uniformitarian logic. Thus ancient calcitic ooids were once all considered secondary, resulting

---

Fig. 3.13 (*opposite*) Carbonate grains in thin section. (a) and (b) Molluscan fragments with well-developed algal borings viewed (a) in plane-polarized light and (b) under crossed nicols. (c) Molluscan fragment with well-developed micrite envelope formed by coalesced and infilled algal borings. (d) Compound intraclast comprising large amorphous lump and micrite-cemented peloids. (e) and (f) Modern ooids viewed (e) in plane-polarized light and

(f) under crossed nicols. Note superficial ooid with pellet nucleus (SO) and the pseudo-uniaxial crosses. (g) Close-up of ooid to show fine laminations and darker areas of organic mucilage and algal borings. (h) Ancient superficial ooid with pelletal nucleus and radial fabric in the ooid cortex. All scale bars 100 μm. ((a)–(c), (e)–(g) from Recent carbonate sands of Bimini lagoon, Bahamas. (d), (h) from Dinantian of Northumberland basin, England.)

from the layer-by-layer replacement of tangential aragonite by radial calcite in concentric organic templates (Shearman *et al.*, 1970). More recently they have been considered as high-magnesian calcite forms (Richter, 1983). By analogy with other similar replacements (e.g. echinoderm carbonate), their fine preservation is ascribed to diffusional substitutions of Fe^{2+} for Mg^{2+} during conversion to low-magnesian calcite on shallow burial.

3.6 Carbonate grains from plants and animals

Whenever we examine a fossiliferous limestone, be it an ancient coral reef community or a fine-grained chalk rich in coccoliths, we are observing a graveyard of former life. Those remains of dead creatures tell us that massive fixation of calcium carbonate has occurred from the world's ocean. The living biomass of carbon is a very tiny fraction of that preserved in the rock record as limestones, coal, oil and natural gas. The living and recently dead biomass provides the main ions needed for chemical weathering and soil formation, the products of which are enfolded into the oceans as nutrients to provide for further growth that utilizes calcium carbonate and silica for metabolism into hard parts and mineralized tissue. Plate 1 shows that marine carbonate sediments are very widely distributed, with the majority derived from the calcareous hard parts of algae and invertebrates.

A local biotic community gives rise to a characteristic death assemblage of calcareous debris (Fig. 3.12). The resemblance between this death assemblage and the original faunal community will obviously reflect the degree of *post mortem* physical, chemical and biological destruction and redistribution. This is the province of palaeoecology and ecological energetics. Many assemblages produce *in situ* carbonate deposits, either as solid reefal build-ups, chiefly of coral, or by the baffling, binding or trapping activities of seagrasses, benthic algae and other organisms. These give rise to unlithified mounds made up of fine-grained carbonate debris and various skeletal remains (see Chapter 24).

Three groupings of marine carbonates may be distinguished:

1 *Oceanic carbonates of pelagic origin.* These are widely distributed in the oceans (with the exception of the North Pacific, Arctic and Antarctic), closely associated with the mid-ocean-ridge system and areas of oceanic deep-water upwelling (Plate 1). They are calcareous oozes comprising the remains of coccoliths, foraminifera (forams) and pteropods. All these originate in the warm shallow photic zones of the ocean and accumulate after death in maximum water depths of 3.5–5 km, as determined by the local carbonate compensation depth (CCD; see Section 3.10).

2 *Shelf carbonates of subtropical and tropical origin.* These are shallow-water carbonates of biological origin. Inorganically precipitated $CaCO_3$ is important locally, as discussed previously. Coral, algal and molluscan species dominate, with coral reefs alone accounting for over 75% of Holocene shallow-water biological $CaCO_3$ fixation (Opdyke & Walker, 1992). The major controls upon organisms that secrete $CaCO_3$ are water depth, temperature and salinity. Water depth is a variable because of the attenuation of life-giving energetic photons in visible light, many of which are lost within the upper 30 m of the water column. Mean sea-surface temperatures of > 15 °C and normal salinities in the range 32–40‰ encourage the so-called 'chlorozoan' assemblage of corals, calcareous green algae and many other phyla. Elevated salinities cause the corals to disappear, defining the 'chloralgal' group.

3 *Shelf carbonates of temperate origin.* These are dominated by contributions from molluscs, forams, red algae and bryozoans, the so-called 'foramol' assemblage. This is more widespread than is usually realized, although accumulation rates are slow because of lowered metabolic rates and lower levels of carbonate oversaturation in cool temperate seas (Burton & Walter, 1987; Opdyke & Wilkinson, 1993).

The ease of identification of calcareous hard parts depends upon the degree of their physical and chemical breakdown. If the sediment is partly or wholly lithified then thin-section analysis techniques must be used. Many complex 3D organisms are difficult to identify in this way: consider the number of possible shapes that a thick-walled cylinder (analogous to a crinoid ossicle) might show if randomly sectioned. Many particles may be classified only to the order level of their appropriate phylum (Fig. 3.14). Particular problems arise if the shell was originally aragonite, since diagenetic dissolution and calcite reprecipitation may totally destroy the original shell structure.

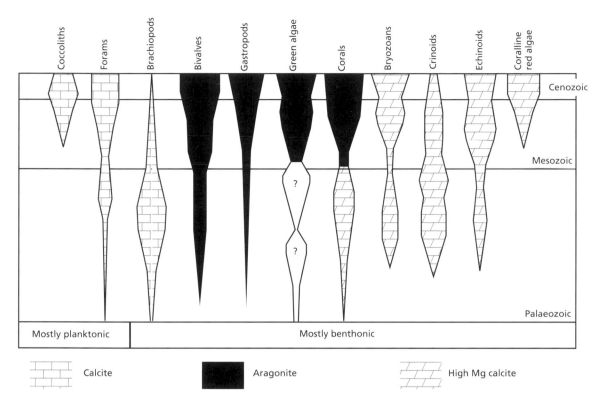

Fig. 3.14 Variation of carbonate mineralogy and phyletic abundance with time for carbonate sediment-forming organisms. The widths of the bars represent relative importance with time of individual *groups* as sediment contributors (not absolute masses of CaCO₃ produced—modern carbonate production is dominated by the humble coccolith). (After Lowenstam, 1963.)

Organisms may secrete their calcareous hard parts as either aragonite, low-magnesian calcite or high-magnesian calcite. Some have multimineral shells. The differing chemical and mineralogical composition of carbonate hard parts clearly gives rise to a corresponding chemical/mineralogical preservation potential. The nature of skeletal carbonate particles

Further information on details of biological grain types

Mollusc shells comprise complex alternations of aragonite and calcite layers arranged in various geometrical ways within protein frameworks and with several characteristic microstructural types (see Bathurst, 1975; Scoffin, 1987; Tucker & Wright, 1990). Most molluscs are benthic creatures, but vast numbers of the pelagic gastropod group, the pteropods, may live in highly productive zones associated with ocean currents or gyres upwelling along shelf ramps (e.g. Gardulski *et al.*, 1990).

Brachiopod shells comprise wholly calcitic thin primary and thick secondary layers composed of fibres. A third inner prismatic layer may occur. Canals crossing the secondary layer may also occur in some groups. Tissue-filled canals normal to the shell wall (filled with calcite cement in fossils) are termed endopunctae. Vertical cone-like deflections

continued on p. 58

of secondary layer fibres are pseudopunctae. Spines occur abundantly on some genera and in certain orientations the careless observer may initially confuse them with ooliths in thin section.

Coral skeletons, including epitheca, tabulae, septae, etc., are made up of tiny calcitic (tabulata, rugosa) or aragonitic (hexacorals) fibres with various orientations.

Foraminiferal tests (Fig. 3.12b) are usually calcitic and they show agglutinating, microgranular, porcellanous, radial fibrous, spiculate and monocrystalline structures. These wall morphologies are, in fact, the primary basis for foraminiferal classification at the sub-order level. Forams may be either benthic or pelagic and make up most of the constituents of oceanic calcareous oozes.

Ostracod shells are calcitic and have a high preservation potential, particularly in deep-sea sediments where they are the only multicellular organisms to have left an abundant fossil record, and hence are valuable palaeoproductivity indicators (Cronin & Raymo, 1997).

Echinoderm debris of high-magnesian calcite mineralogy comprises fenestral elements with a very characteristic porous structure, individual particles behaving as single calcite crystals under crossed polars. Rapid diagenetic diffusional magnesium-loss and pore infilling by low-magnesian calcite cement still preserves this single-crystal behaviour, so that echinoderm debris in limestones may, with practice, be recognized at a glance under crossed polars. Original echinoderm high-magnesian calcite is only rarely preserved in the geological record (Dickson, 1995).

Calcareous benthonic algae show diverse structures. Erect frond-like forms such as the green alga *Halimeda* and the red alga *Lithothamnion*, the latter most important in temperate carbonates, break down initially into gravel-sized segments, the tubular internal structure of the former being especially characteristic. Many algae show a cellular internal structure. Calcification may occur within the individual cells, in the cell wall or outside the cell wall. After death, codiacean algae (e.g. *Penicillus*; Fig. 3.15) may break down completely into needles of aragonite that may contribute much of the aragonite mud found in low-energy shallow environments in tropical or subtropical areas. Filamentous calcareous algae such as *Girvanella*, *Garwoodia* and *Ortonella* are important encrusting agents in many ancient limestones, particularly Upper Palaeozoic and Mesozoic.

Calcareous planktonic algae of the phylum Chrysophyta ('golden' algae) are abundantly represented in many pelagic deposits from the Jurassic onwards. They are referred to broadly as calcareous nanoplankton, important amongst the primary producers of the oceanic surface waters. Originally, in life, the tiny circular to oval discs or scales that we know as coccoliths surrounded a single algal cell up to about 80 μm in diameter. *Emiliania huxleyi* exemplifies the Corinthian beauty of these exquisite organisms (Honjo, 1976). Coccoliths thrive in the sunlit waters of the ocean surface where nutrient concentrations are high. The calcitic coccolith plates are typically 2–20 μm in maximum diameter. Their seabed preservation depends entirely upon the depth of the CCD (Section 3.10).

Further information on details of biological grain types [*continued*]

Fig. 3.15 Calcareous algae of the genus *Penicillus* with holdfasts. Vast numbers of these algae live in the shallow lagoons of Florida and the Bahamas. Upon death, the delicate organic tissues decay and release tiny, blunt-ended aragonite needles of the plant 'skeleton' on to the lagoonal floor. Some proportion accumulates as aragonitic mud indistinguishable in external form from precipitated aragonite.

formed in a specific environment (e.g. reef, intertidal flat, shelf sand) has obviously changed through geological time in response to evolution. Thus a Silurian reef faunal and floral assemblage is distinct from a Carboniferous, Triassic or modern assemblage.

3.7 Carbonate muds, oozes and chalks

Regardless of the controversial chemical results concerning aragonite precipitation discussed previously, it is apparent that the net accumulation of tiny aragonite needles, chiefly from the breakdown of benthonic calcareous algae (Fig. 3.15), can contribute significantly to known aragonite deposition rates in many areas of the Bahamas–Florida carbonate province (Stockman *et al.*, 1967; Bosence *et al.*, 1985). It is also known that the extensive 'aragonite' muds west of Andros Island (Great Bahamas Bank) contain a relatively large proportion of high-magnesian calcite, which suggests a source from the breakdown of red algae for some proportion of these muds.

An ooze is defined as a deep-water deposit of predominantly pelagic origins comprising > 50% sub-sand-sized carbonate debris. Sand-sized organic grains are dominated by forams and pteropods. Studies on shelf ramps indicate the importance of ocean current dynamics in determining the relative proportion of these two grain types. Pteropods thrive in fertile waters of more-or-less constant (stenohaline) salinity, whilst some forams can tolerate wider salinity fluctuations. Deposits cored on the Florida ramp reveal alternating pteropod-rich and foram-rich layers whose cyclicity is attributed to interglacial and glacial periods respectively (Gardulski *et al.*, 1990). Data from oceanic oozes indicate that during glacial periods calcitic pelagic organisms are replaced by siliceous plankton in regions of lower water temperatures and high productivity, with the silica–calcite 'front' moving generally equatorwards and the proportion of $CaCO_3$ preserved in bottom sediments decreasing and that of organic carbon increasing (Archer & Maier-Reimer, 1994).

The accumulation of coccolith debris above the CCD in the oceans is possible because of the fantastic $CaCO_3$ production rates of these plants. In fact the coccolith oozes of the shallow mid-ocean ridges are one of the largest sinks for $CaCO_3$ on Earth, Plate 1. It has been observed that individual coccolith plates are secreted and extruded through the surface of the coccolith algal cell once every 2 h in summer sunlight. The results are gigantic 'blooms' of coccoliths, which are visible from aircraft and satellites.

3.8 Other carbonate grains of biological origins

Micrite envelopes

Microscopical examination of skeletal fragments collected from quiet-water carbonate environments often shows that a dark rind surrounds their exteriors (Fig. 3.13). Closer investigation shows that the rind is of variable thickness and intensity around different fragments. Individual tube-like cavities may be seen in the less advanced rinds and on the inner margins of the thicker rinds. The impression gained is that the rinds are formed by the coalescence of these cavities. The tubes may be empty or they may be filled with very fine-grained aragonite or high-magnesian calcite. Dissolution of the shell-fragment carbonate with dilute acid yields a gelatinous residue in which various types of blue-green algae may be identified. The tubes have thus been postulated to be the result of boring blue-green algae (fungal bores also exist), which rapidly infest all carbonate particles after deposition (Bathurst, 1966). The reason for carbonate precipitation in the bores is poorly known but may be connected with the local high pH in the tube following CO_2 uptake by adjacent photosynthesizing algae or by bacterial action on the organic residue left in vacated tubes. Algal bores found in temperate-water carbonates do not show carbonate precipitation (Gunatilaka, 1976).

Micrite envelopes produced by coatings of filamentous endolithic algae are described by Kobluk and Risk (1977). Rapid precipitation of low-magnesian calcite occurs on and within dead filaments, which project from the substrate into the sea. Coalescence of the dead calcified filaments could produce an algal micrite envelope that is accreted on to a grain periphery previously bored and micritized by Bathurst's mechanism described above.

Amorphous lumps and intraclasts

Micrite envelopes formed by boring algae may extend inwards to include the whole shell fragment. An amorphous lump particle is thus produced, with no tell-tale shell structure left to testify to its original shell-fragment origin (Bathurst, 1966). A high proportion of such angular lumps make up the sediment in certain Bahamian and Persian Gulf lagoons. Such

particles must, in fact, be defined as belonging to the class of carbonate particles known as intraclasts. The class is usually defined as carbonate particles that have been reworked into the basin of deposition (Folk, 1962). Micritic amorphous lumps formed by micritization are *not* true intraclasts but *must* be included in the term since they are indistinguishable from true reworked intraclasts from other sources. Intraclasts are thus a highly diverse, polygenetic group of particles, including reworked beachrock fragments, hardground debris, grapestones, older lithified carbonate particles and amorphous lumps of skeletal origin.

Faecal pellets

In quiet-water lagoons with aragonitic mud substrates, large numbers of molluscs (chiefly gastropods and bivalves), worms and crustaceans continually ingest organic-rich muds, feeding upon the nutrients. The mud is excreted as ovoid faecal pellets. Internally the pellets are dark, fine-grained, usually structureless and rich in organic matter. Older pellets are fairly hard, probably due to rapid interstitial carbonate precipitation. Winnowing of associated aragonite mud may cause a sand-grade deposit of pellets to be produced, as in lagoons of the Arabian Gulf.

Faecal pellets in ancient carbonate rocks may be very difficult to tell apart from abraded amorphous lumps produced by micritization of shell fragments and other processes, including complete micritization of ooliths. For this reason it is best to use the term *peloid* for any structureless ovoid micrite particle unless the genesis may be deduced by some other observations.

Grain aggregates—grapestones

Large areas of the Bahamas Banks are covered with composite particles that comprise cemented aggregates of grains (shell fragments, ooids, pellets) resembling microscopic bunches of grapes (Fig. 3.12c). Intense micritization by infilled algal bores usually obscures any original internal structure of the particles. It is thought that the cementation, by micritic aragonite, and the intense micritization reflect a mode of growth within a subtidal blue-green algal mat (see Chapter 24) that stabilizes the substrate to resist tidal and wave currents. Periodic mat rip-up during storms

then yields grapestone aggregates. Grapestones are thus clearly a type of intraclast, as defined previously.

Algaliths

Algaliths may be defined as unattached nodules (oncoids) of carbonate-secreting algae. Rhodoliths (Bosence, 1983a,b) are nodules of red coralline algae, and are widely distributed from the tropics to the poles and are a particularly important contributor to temperate carbonates. They comprise laminar crusts, branching or columnar/mamillated forms, the exact morphology depending to a strong degree on energy conditions. Cyanoliths (Riding, 1983) are nodules constructed by cyanophytes, whose beautifully delicate sheath-like cells are preserved by calcite precipitation.

3.9 Organic productivity, sea-level and atmospheric controls of biogenic $CaCO_3$ deposition rates

Direct investigations of productivity amongst standing crops of the algae *Penicillus*, *Halimeda*, *Neogoniolithon* and other mud contributors (like the epifaunas attached to the sea-grass *Thalassia* in Floridan mud mounds) indicate that about 500 g/m² of $CaCO_3$ biomass is added per year. This is equivalent to a deposition rate of about 2 mm/yr. It is instructive to convert this to carbon productivity (multiplying by 0.1 to obtain) and to compare this with primary organic productivity in the same area (data in Pernetta, 1994). The production rates are of the same order of magnitude. Sediment budget studies suggest that algal sediment (and chemically precipitated aragonite) is overproduced compared to observed depositional fluxes and that very substantial loss of aragonite occurs to the shelf as suspended plumes, probably during storms and hurricanes (Neumann & Land, 1975; Bosence *et al.*, 1985).

Coral reef growth can vastly exceed lagoonal or forereef rates of deposition. Extraordinary rates of up to 6 mm/yr have been observed. Interestingly enough the typical reef-building coral *Acropora* has its most rapid rates of extension at night, with the algal symbiont zooxanthellae playing little role in this aspect of growth (Vago *et al.*, 1997). Darwin recognized the importance of high coral growth rates over 150 years ago and featured it in his novel 'volcano-drowning' hypothesis for the origin of atolls. Coral reefs are by far the greatest sink for $CaCO_3$ in the modern oceans and they have been assigned a key role in the recycling of carbon during interglacial as compared with glacial intervals (Berger, 1982; Opdyke & Walker, 1992; Archer & Maier-Reimer, 1994). Highstand interglacial periods have very high 'coral reef' carbonate fixation and relatively high atmospheric CO_2 concentrations. The opposite holds for glacial lowstands. Thus the modern (highstand) fixation of $CaCO_3$ by the world's coral reefs cannot be much less than about 2×10^{13} mol/yr for the last 5000 yr. This is approaching the total Ca^{2+} flux into the ocean of about 2.4 mol/yr. During lowstands the shallow shelf area available for coral reef colonization is much reduced and thus $CaCO_3$ fixation with it. During such times the locus of carbonate deposition must shift to the deep sea (Plate 1). Further, glacial lowstands will show marked dissolution of highstand reefs and other porous carbonates, leading to a net seaward alkalinity flux. Model results indicate that this highstand–lowstand cycle is capable of predicting a significant proportion of atmospheric CO_2 compositional shifts as measured in fossil ice cores. There is an important additional contribution due to seafloor and shallow-subsurface CO_2 production by organic carbon oxidation (Archer & Maier-Reimer, 1994).

As noted previously, temperate-water carbonates of the foramol assemblage lack the oolith, intraclast and pellet allochems and the contribution from green algae and corals that characterizes warm-water carbonates (Lees, 1975). Their deposition rates are very much lower, as a consequence of the rapid decrease in metabolism and carbonate supersaturation with increasing latitude (Fig. 3.16; Opdyke & Wilkinson, 1993). Algal bores in temperate carbonates appear to remain unfilled; therefore micrite envelopes should be uncommon, a point that has some relevance to the preservation of skeletal aragonite in such carbonates.

3.10 $CaCO_3$ dissolution in the deep ocean and the oceanic $CaCO_3$ compensation mechanism

The world map of the distribution of $CaCO_3$ deposits (Plate 1) shows a remarkable coincidence between the crests and flanks of mid-ocean-ridge systems and the occurrence of pelagic calcareous oozes formed by coccoliths and forams. Detailed mapping shows that carbonate sediments are rare below a water depth

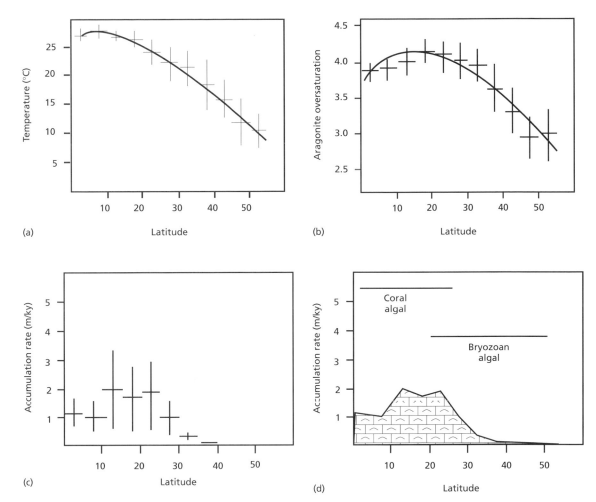

Fig. 3.16 (a) and (b) Sea-surface temperature and aragonite saturation as a function of latitude. Trends defined by best-fitting second-order polynomials drawn through over 200 data points (individual points not shown for clarity). Vertical error bars are for one standard deviation drawn every 5° of latitude (horizontal lines).

(c) and (d) Holocene carbonate deposition rate data (> 200 data points), summarized as mean values every 5° of latitude and one standard deviation. Also shown are the major subtropical and temperate facies groups. (All after Opdyke & Wilkinson, 1993.)

of about 5 km in the Atlantic and about 3.5 km in the Pacific. This distribution depends partly upon the increased solubility of $CaCO_3$ in the deep oceans. It has been calculated that about 80% of the $CaCO_3$ produced in the warm sunlit areas of ocean waters by planktonic organisms is destroyed by deep-water dissolution. Some of the carbonate that had survived oceanic descent is dissolved in sediment porewaters. Direct measurements and modelling studies indicate that about 30–50% of deposited calcite is subsequently dissolved in this environment (Archer, 1991).

The temperature of oceanic water decreases very rapidly to about 5 °C at around 1000 m and thereafter more slowly to reach a minimum of around 2 °C. These cold deep waters contain more CO_2 in solution than do warm surface waters. More important is the increase of pressure with water depth, which also causes increased p_{CO_2} and thus lower pH. Both effects cause oversaturation of seawater with respect to $CaCO_3$ to decrease with increasing depth. There is also another effect crucial to the deep oceanic acidity–alkalinity balance (Archer & Maier-Reimer, 1994).

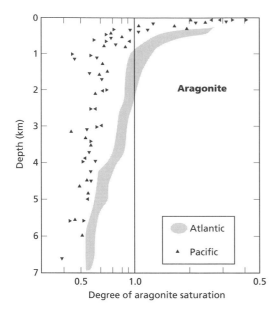

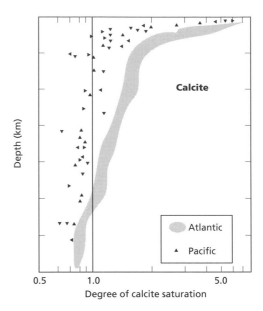

Fig. 3.17 The degree of aragonite and calcite saturation as a function of depth in the Atlantic and Pacific oceans. (Data of Broecker, 1974.)

This is the rate of production of total organic carbon in the oceanic mass and the proportion of its buried flux that is oxidized by respiration (producing CO_2) either at the ocean floor or in the few tens of metres below the sediment–water interface. During glacial periods the production of organic carbon increases: the increased degradation of this produces more CO_2, which dissolves more sedimented $CaCO_3$, which drives the various $CaCO_3$ compensation reactions discussed above to increase total ocean alkalinity.

Aragonite, the more soluble $CaCO_3$ polymorph, becomes undersaturated long before calcite, at about 500 m depth in the Pacific ocean and 2 km in the Atlantic ocean (Fig. 3.17). Thus pelagic organisms with aragonite skeletons, like the pteropods (a group of tiny gastropods), are much more prone to dissolution than are the calcitic coccoliths. Calcite becomes undersaturated at depths of between 400 and 3500 m in the Pacific and between 4000 and 5000 m in the Atlantic. These zones where $CaCO_3$ approaches undersaturation for aragonite or calcite are known as carbonate saturation depths (CSD). Whether or not calcitic organisms can survive as debris below the CSD will be a complex function of the rate of fall of the grain (or agglomeration of grains in faecal pellets)

through the water column vs. the rate of $CaCO_3$ dissolution. The rate of dissolution thus depends partly on skeletal size. Solution inhibition by adsorbed molecules may also be important (see Section 3.3).

The variation of CSD between oceans is largely a function of pH. Pacific ocean water contains much larger quantities of organic debris than does Atlantic ocean water. Oxidation of this debris produces additional H^+ ions, some of which are used up in seawater buffer reactions; the small remainder not balanced in this way increase the deep-water acidity. Calcium carbonate is thus dissolved and is less abundant in Pacific than in Atlantic sediments. Adsorbed phosphate and organic ions also act as inhibitors to dissolution (Chave & Suess, 1970), and it is not until these are removed in the deep ocean that dissolution can proceed rapidly (Morse & Berner, 1972; Takahashi, 1975; Berner, 1976).

From the above discussion it should be clear that the oceans are highly sensitive to changes in sources and sinks for $CaCO_3$. Any change in the global balance of weathering and alteration (input) or shallow- and deep-water deposition (output) will change the oceanic carbonate-ion content so that a balance is restored. This is the $CaCO_3$ compensation mechanism and may be written as the following notional pH balance:

$$CO_2 + CO_3^{2-} + H_2O \rightleftharpoons 2HCO_3^- \qquad (3.4)$$

Mineral	Formula	IAP	K
Halite	NaCl	0.12	38
Gypsum	$CaSO_4 \cdot 2H_2O$	4.6×10^{-6}	2.5×10^{-5}
Anhydrite	$CaSO_4$	4.6×10^{-6}	4.2×10^{-5}
Sylvite	KCl		
Kieserite	$MgSO_4 \cdot H_2O$		
Carnallite	$KMgCl_3 \cdot 6H_2O$		
Polyhalite	$K_2MgCa_2(SO_4)_4 \cdot 2H_2O$		

Table 3.3 Chemical composition and selected values of IAP (ion activity product) and K (solubility product) for halite, gypsum, anhydrite and other evaporites in seawater solutions. (After Berner, 1971.)

The reaction maintains pH equilibrium by an inverse relationship between the first two species on the left-hand side.

Palaeoceanographic studies indicate a major deepening of the CCD, and thus an increase in amount and extent of coccolith calcite at the Eocene/Oligocene boundary (van Andel, 1975), coincident with the initiation of major Antarctic ice cover. This might be thought to be the opposite effect than would be expected from enhanced oceanic circulation and cooler bottom-water production following ice-sheet growth and global cooling. That is, until the realization that modern cool North Atlantic Deep Water is much less corrosive to calcite than the more acidic, organic-rich Pacific waters. The problem is a complex one. More efficient circulation would also have encouraged higher primary oceanic productivity because of greater nutrient fluxes. The resulting higher organic turnover in the surface would have been accompanied by greater organic decay at depth, leading to lower global oceanic pH and hence a shallower CCD. Possible additional compensating factors are a decrease in atmospheric CO_2 and an increase in the continental Ca flux. There is good evidence for greatly increased continental weathering recorded in the Tertiary Sr record (see Section 15.8). Increased Ca would have 'mopped up' excess acidity due to higher productivity and perhaps overcome Ca-limited growth of calcareous plankton.

3.11 Evaporite salts and their inorganic precipitation

Evaporitic salts are precipitated from concentrated natural brines in basins with limited connections to the open ocean, or in closed continental basins. The high evaporation rates required for this process of concentration involve both high temperatures and low relative humidity. Thus even in the face of high surface temperatures, if sea winds are saturated, little evaporation will result (see discussion of Kinsman, 1976). Holocene environments have not created the necessary conditions for large-scale marine evaporite precipitation, although several periods during the Phanerozoic have been marked by gigantic precipitation events. The opening of the Atlantic and the Red Sea ocean basins, in the early Mesozoic and mid-Cainozoic respectively, are good examples, as is the incomparable story of the desiccation of much of the Miocene Mediterranean (Hsu, 1972).

The main evaporite mineral phases are listed in Table 3.3. In this section we shall concentrate upon marine evaporites precipitated from standing brine bodies of marine origin. Further considerations of diagenetic evaporites found in sabkhas and evaporite facies can be found in Chapter 24.

Normal seawater is undersaturated with respect to all evaporitic salts (Table 3.3). Note that halite is much more undersaturated than gypsum or anhydrite, so that any evaporation will cause gypsum to precipitate before halite. The most undersaturated salts are the complex series of potassium salts and these represent the final precipitates from highly concentrated brines. It should be noted that progressive evaporation is not a linear process since it is progressively more difficult to vaporize water from brines because of changes in surface tension. Figure 3.18 shows schematically the sequence of salts that might precipitate as seawater is progressively concentrated by evaporation.

This sequence differs somewhat from the actual sequences recorded in evaporite deposits in the geological record. The latter show increased proportions of $CaSO_4$ and decreased proportions of sodium/magnesium sulphates compared with laboratory studies. Magnesium depletion occurs by a combination of

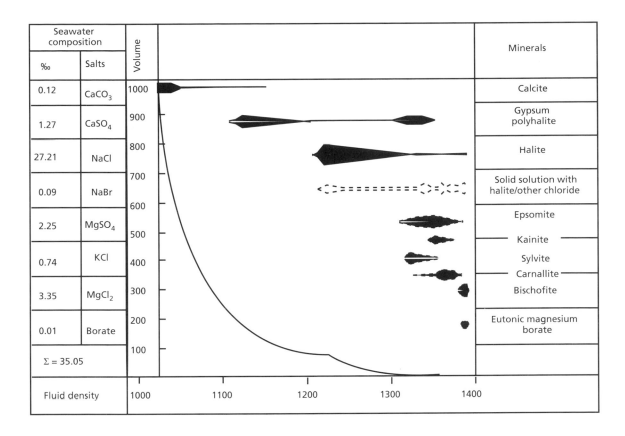

Fig. 3.18 The effects of seawater evaporation upon brine volume, density and type of salt precipitate. (After Valyashko, 1972.)

dolomitization and clay mineral precipitation. Important changes due to the effects of percolating bitterns arise during the last stages of brine concentration. Also, influxes of seawater may cause dissolution and reprecipitation.

Anhydrite has never been observed to precipitate directly from seawater. The equilibrium constant for the reaction

$$CaSO_4(solid) + 2H_2O \rightleftharpoons CaSO_4 \cdot 2H_2O(solid) \quad (3.5)$$
$$\text{anhydrite} \qquad\qquad\qquad \text{gypsum}$$

is given by the activity of water, $a^2_{H_2O}$ (Hardie, 1967), and enables stability fields for gypsum and anhydrite to be plotted from experimental results (Fig. 3.19).

In highly saline brines, gypsum is still precipitated, but as a metastable phase, which may subsequently

alter to anhydrite. This process occurs in sabkha evaporites (Chapter 24), but textural evidence indicates primary precipitation in some of these environments. Experiments (Cody & Hull, 1980) indicate that primary anhydrite precipitation is possible in the presence of certain organic molecules that act to inhibit gypsum precipitation. The kinetics of eqn (3.5) indicate that the increase of pressure due to natural burial will favour the denser calcium sulphate phase, causing gypsum to transform to anhydrite, and vice versa for exhumation.

The geochemistry and composition of continental brines is much more complicated and unpredictable than that of marine brines, basically because there is no all-buffering oceanic reservoir to smooth out contrasts between local ionic inputs. Streams and groundwaters will have compositions determined by amount of runoff, catchment bedrock type and local chemical weathering processes (see Chapter 2), with important contributions from longer-travelled groundwater brines. Eugster and Hardie (1978) and Harvie *et al.*

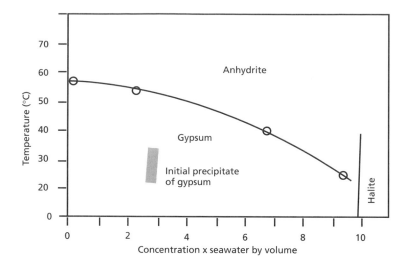

Fig. 3.19 Phase diagram to show the stability fields of gypsum and anhydrite with respect to brine temperature and seawater concentration. (Data of Hardie, 1967, as recalculated in terms of concentrations (from a_{H_2O}) by Blatt *et al.*, 1980.)

(1982) distinguish seven major water types in continental brines (Fig. 3.20): those poor in alkaline-earth elements, i.e. (a) $Na–CO_3–Cl$, (b) $Na–CO_3–SO_4–Cl$; those poor in calcium, i.e. (c) $Na–SO_4–Cl$, (d) $Na–Mg–SO_4–Cl$, (e) $Mg–SO_4–Cl$; and those poor in sulphate, i.e. (f) $Ca–Na–Cl$, (g) $Ca–Mg–Na–Cl$.

Marine brines approximate to type (d), whilst many subsurface connate brines approximate to type (g). Brine composition obviously depends largely upon mineral precipitation, and a simplified brine evolution scheme is summarized schematically in Fig. 3.20. This shows the two solid chemical 'divides' due to gypsum and dolomite precipitation. Continental evaporites are distinctive in that HCO_3^- and CO_3^{2-} dominate the anions and are still available in the late stages of fractionation to form alkaline-earth carbonates (like trona–a complex sodium carbonate). In seawater, calcium and carbonate ions are used up early on in calcite, aragonite or dolomite precipitation reactions, leaving fractionated brines poor in Ca^{2+} and CO_3^{2-} and rich in Na^+, Cl^- and SO_4^{2-}. As noted previously, the lack of ancient sodium carbonates in marine evaporite sequences argues strongly for a fairly constant oceanic pH and composition, approximating to the Recent, throughout most of geological time.

It has long been recognized that simple concentration of seawater by evaporation is insufficient to produce the great thicknesses of evaporite salts observed in the geological record. For example, it may be calculated that complete evaporation of the world's oceans will yield a mean thickness of only 60 m of evaporites. Some ancient evaporite successions of great areal extent may reach over 1 km in thickness. Major periods of evaporite production must have had drastic short-term effects upon levels of seawater salinity and oxygen isotope composition before a steady state was once more established. Thus the total volume of salts in the world's oceans is about 2.2×10^7 km³ (Borchert & Muir, 1964). The Permian Zechstein evaporites of northwest Europe have a total volume of about 2.4×10^6 km³, some 10% of the oceanic reservoir! However, spectacular short-term events such as those mentioned must be set aside the long ($> 10^8$ yr) mean residence time of chloride in the oceans. It has been calculated that the total mass of Cl^- in sedimentary rocks and porewaters is roughly equal to that of the modern oceans, and so it is unlikely that the salinity of the oceans could have changed by more than, say, 50% above its present value since the early Precambrian.

Models for evaporite evolution must obey a hydrological accounting system and have the following components: initial brine mass, rate of influx from 'mother' basin, rate of evaporative loss, rate of evaporite solid deposition and rate of seepage or outflow (termed reflux) of dense brine (Fig. 3.21). Clearly the system is a complicated one, even neglecting dynamic effects arising from internal stratification and convection. Some light may be shed on the vexed problem of the origin and depth of evaporating brines by making use of bromine geochemistry. This element

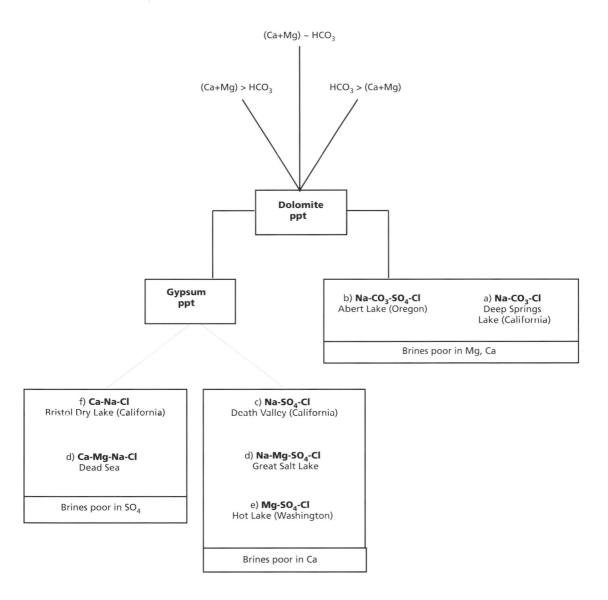

Fig. 3.20 Flow diagram to show brine types and evolution. (After Harvie *et al.*, 1982.)

does not precipitate as a Br salt, but is incorporated in the halite lattice. Halites precipitated from marine-sourced brines have much greater Br contents than halites from non-marine-sourced brines; a value of less than 65 ppm Br is usually taken to be the upper limit for the latter case. The partition coefficient of Br⁻ between crystallizing halite and the residual brine is around 0.1, leading to increased Br in halite as crystallization proceeds (Valyashko, 1972). It is possible simply to model the rate of change of bromine with height in a halite deposit precipitated from a closed basin, but estimates of brine depth are very large. Tucker and Cann (1986) show that, once open system conditions are realistically modelled (as functions of rates of influx and reflux), more realistic depths are found, e.g. 140 m instead of 2665 m for an analysed Zechstein profile.

The simplest model for subaqueous evaporite formation is the shallow-water barred basin (Fig. 3.21) where evaporation proceeds in semi-isolation, with replenishment of seawater over a restrictive entrance

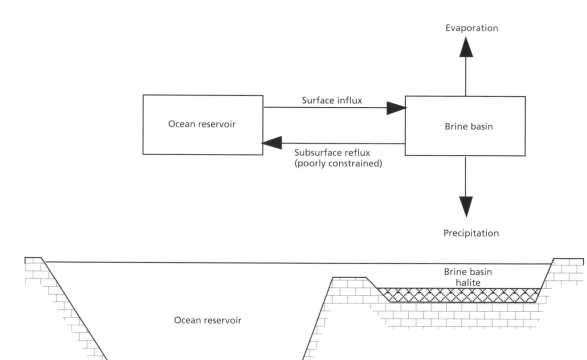

Fig. 3.21 A simple input–output model for a brine basin. (Inspired by Tucker & Cann, 1986.)

sill or obstruction. Brines are thus progressively concentrated, and progressive crystallization of the various salts then proceeds. In such models, depending upon the rates involved, basin brines may stay at particular concentrations for long periods. In this way abnormally thick sequences of sulphates or chlorides may accumulate. Cycles of evaporites that approximate reasonably closely to the 'ideal' cycle result when basin brines are evaporated to completion. Lateral changes in evaporite composition are to be expected in barred basins since incoming seawater will precipitate first gypsum and then halite as it spreads over the sill towards the basin shallows. Subaqueous evaporite facies are further discussed in Chapter 24.

One further point concerns the presence of small-scale evaporite rhythms on a millimetre to centimetre scale observed in many ancient evaporites (Fig. 3.22).

These may comprise alternating (a) dolomite–anhydrite + clastic clay, (b) clay dolomite–anhydrite + halite, and (c) halite + sylvite–carnallite. The clay interlaminae in types (a) and (b) are interpreted as a clastic influx during rainy seasons. The inflowing waters 'freshen' the brine body and this, together with cooler air temperatures, causes either cessation of evaporite precipitation or precipitation of a less undersaturated phase. The laminae thus represent annual varves by this interpretation (Richter-Bernberg, 1955). It is not unfeasible that some layers may also be the product of double diffusive convection, analogous to that thought to occur in magma chambers and reproduced in simple analogue experiments.

We must finally mention the apparent occurrence of global cycles in marine evaporite deposits, perhaps reflecting secular variations in seawater chemistry. Hardie (1996) has noticed that large marine potash evaporite bodies are dominated either by Mg-rich phases like polyhalite and kieserite or by KCl salts like sylvite. He points out that the periodicity of these

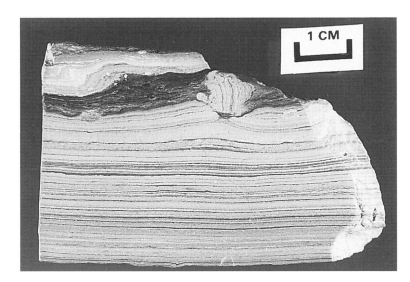

Fig. 3.22 Subaqueous evaporite precipitates: varved anhydrite–dolomite clay from the late Jurassic Hith Formation of Saudi Arabia. Note reworked clast at top of specimen.

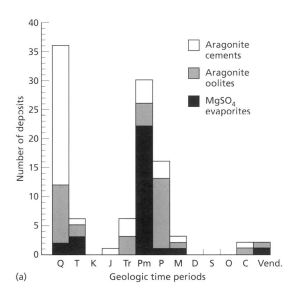

(a)

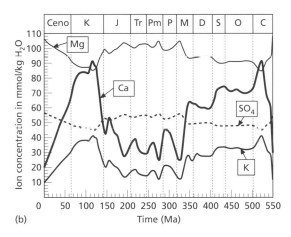

(b)

Fig. 3.23 (a) Time changes in carbonate and evaporite mineralogy. (b) Time changes modelled for variation in major ion chemical composition of the Phanerozoic oceans. (Both from Hardie, 1996.)

trends coincides with those established for carbonate minerals, viz. the occurrence of aragonite vs. calcite seas. In particular, it seems that aragonite seas coincide with periods dominated by magnesium sulphate evaporites (Fig. 3.23) and calcite seas with KCl evaporites. The explanation is sought in large-scale secular seawater compositional changes due to changing rates of seafloor spreading causing mid-ocean-ridge hydrothermal/submarine weathering reactions to provide greater or lesser amounts of magnesium, calcium, potassium and sulphate to the oceans.

3.12 Silica and pelagic plankton

Seawater is vastly undersaturated with respect to amorphous silica. Thus for the system:

$$SiO_2 + 2H_2O \rightarrow H_4SiO_4 \tag{3.6}$$

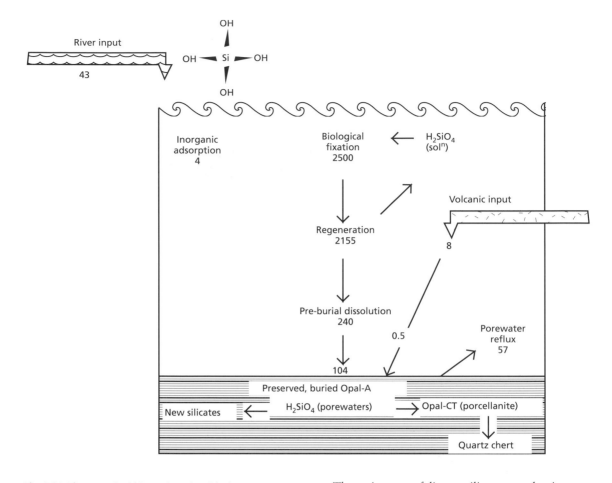

Fig. 3.24 The oceanic Si biogenic cycle with the magnitudes of dissolved silica as SiO_2 in 10^{13} g/yr. Note the dominance of biological fixation and regeneration. Annual river input is roughly balanced by diatom burial. (Data of Heath, 1974; Riech & von Rad, 1979.)

$K = a_{H_4SiO_4} = 2 \times 10^{-3}$ and IAP is 2×10^{-4} to 1×10^{-6}. Most dissolved silicon occurs in the H_4SiO_4 form, further ionization to $H_3SiO_4^-$ being very limited ($K = 10^{-9.9}$). The undersaturation arises because of highly efficient removal of Si by siliceous plankton, of diatoms and radiolarians, which construct their skeletons with the amorphous opal-A form. This biogenic extraction is opposed by important bacterially-mediated dissolution of skeletal material at depth (see Biddle & Acam, 1999), causing a rapid increase of dissolved silica with depth in the oceans. The total deposition of biogenic silica per year is about 10^{13} mol (Drever *et al.*, 1988).

The main areas of diatom siliceous production correspond to areas of high surface productivity where ocean current divergence or upwelling causes fertile deep waters rich in phosphorus, nitrogen and iron to rise into the warm photic zones (see Chapter 26). The radiolaria are not primary producers, however (i.e. they do not photosynthesize), and feed upon the abundant remains of various primary producers. We have already discussed data from oceanic oozes that indicate that during glacial periods calcitic pelagic organisms are replaced by siliceous plankton in regions of lower water temperatures and high productivity. The silica–calcite 'front' moves generally equatorwards, the proportion of $CaCO_3$ preserved in bottom sediments decreases and that of organic carbon increases (Archer & Maier-Reimer, 1994). An outline of the silica cycle in the oceans is shown in Fig. 3.24.

Siliceous diatoms, algal primary producers, are also

Fig. 3.25 Freshwater diatoms from middle Holocene sediments of proto-Lake Hebgen, Montana, USA. Scale bar = 10 μm. (Photo courtesy of J. Alexander.)

a feature of many temperate lake deposits (Fig. 3.25) where they also cause marked near-surface depletion in dissolved silica as the diatom tests are sedimented faster than the dissolved silicon is replaced by runoff (Wetzel, 1983, pp. 333–8). Typically, maximum diatom productivity in temperate lakes occurs at the time of the early spring overturn when the surface waters have their maximum concentrations of nutrients. Important fixation of Si by diatoms also occurs in the highly productive flooded wetlands (Konhauser et al., 1993) and estuaries of major tropical rivers, notably the Amazon.

The nature of the Si cycle in the Precambrian is of some interest (see Siever, 1992) because radiolarians are first recorded during the Cambrian and the more efficient diatom silica users date from as late as the Cretaceous. Because of the lack of such efficient organic Si extractors, it is likely that the Precambrian oceans had much more dissolved Si than today. Pore-water diffusion, Si exchange and adsorption with clay minerals are the likely methods of inorganic early diagenetic silica precipitation. Bedded cherts of apparent inorganic origins are widely recorded, particularly as fine interbeds with haematite in the enigmatic Proterozoic banded ironstone formations (BIFs) (see next section).

3.13 Iron minerals and biomineralizers

Five physicochemical variables control the nature of reactions involving Fe^{2+} and Fe^{3+} with other dissolved species. These are E_h, pH and the activities of dissolved HS^-, HCO_3^- and Fe^{2+}. Haematite and other ferric species are the only iron minerals that can exist in equilibrium with depositional waters above the sediment–water interface. Much evidence is gathering that both microbial reduction of Fe^{3+} to Fe^{2+} *and* oxidation of Fe^{2+} to Fe^{3+} are common under anaerobic conditions (Lovely et al., 1987; Widdel et al., 1993; Brown et al., 1997). The Fe^{3+}-reducing microorganism strain GS-15 metabolizes by using enzymes to link the oxidation of natural acetate to CO_2 with the reduction of Fe^{3+} to Fe^{2+}. Thus:

$$CH_3COO^- + 8Fe^{3+} + 4H_2O \\ \rightarrow 8Fe^{2+} + 2HCO_3^- + 9H^+ \tag{3.7}$$

Interestingly, this same reaction, but now involving the reduction of U^{6+} to U^{4+}, yields more energy to the bugs and has been proven to proceed independently (Lovely et al., 1991), providing a new method of mobilizing insoluble U^{6+}.

The natural oxidation of Fe^{2+} to Fe^{3+} may occur by bacteria or by chemical oxidation, both involving molecular oxygen, *or*, as discovered by Widdel et al.

(1993), by a group of purple bacteria present in all marine muds that utilize the process in the absence of oxygen in order to reduce CO_2 to cell material, i.e. the bacteria photosynthesize utilizing energy from the reaction $Fe^{2+} \rightarrow Fe^{3+}$. These anoxygenic phototrophic bacterial strains are very important because they imply that oxygen-independent iron oxidation was possible long before the evolution of oxygenic photosynthesis. As such they may provide a key in our understanding of Precambrian BIFs briefly discussed below, perhaps even obviating the need for abundant free oxygen in their formation.

Another source of minute magnetite grains in sediments (mostly forming in the shallow subsurface sulphate-reducing zone) is the magnetotactic bacteria, in whose cells crystals of magnetite precipitate as part of an intricate and variable biomineralization process (Bazylinski *et al.*, 1993). It is these prokaryotes that impart the majority of depositional remnant magnetization to sediments. Even continental sediments that contain predominantly haematite seem able to preserve this bacterial signature.

Goethite is the major source of detrital iron in sediments. It is derived from weathering and soil reactions, including lateritization. Clay minerals may contain Fe^{2+} in their lattices or as adsorbed species on the clay mineral surfaces (Carroll, 1958). Once deposited, goethite must reach an equilibrium with haematite by the dehydration reaction (Berner, 1969):

$$2HFeO_2 \rightarrow Fe_2O_3 + H_2O \qquad (3.8)$$

$\Delta G°$ for this reaction is always negative, the exact value being dependent upon goethite crystallinity. Thus limonitic goethite is unstable relative to haematite + water under diagenetic conditions. This explains the complete absence of yellow to brown limonite/goethite sediments in the subsurface. Arid-zone weathering encourages red haematitic pigment production with time in near-surface environments. The haematite may remain stable as long as organic matter remains absent. The presence of organic material will encourage reduction of Fe^{3+} to Fe^{2+} and the red pigment will disappear.

Pyrite is a common diagenetic iron mineral in many marine sediments. Below the thin zone of oxygenated porewaters, anaerobic bacterial reduction of sulphate occurs. The H_2S that forms reacts with iron to form the iron monosulphide FeS. Continued H_2S production encourages sulphur-oxidizing bacteria to form elemental sulphur. This sulphur reacts with the FeS

over a period of years (Berner, 1970) to form microscopic aggregates of pyrite crystals termed framboids. The most important factor limiting pyrite formation is the availability of organic matter that can be metabolized by bacteria. Both iron and sulphate species are usually present in abundance in marine diagenetic porewaters. Thus the greater the amount of organic material, the greater the amount of pyrite produced. Marine black shales rich in organic material are frequently pyritous in the geological record. By way of contrast, non-marine diagenetic porewaters are generally very low in dissolved SO_4^{2-} and hence pyrite is usually absent. Rapid marine deposition will tend to inhibit pyrite formation since there will be a limited time available for SO_4^- diffusion from the overlying marine reservoir. The SO_4^{2-} available is restricted to that contained within the buried porewaters.

Siderite will form as a diagenetic mineral only where a very low dissolved sulphide concentration is coupled with high dissolved carbonate, high Fe^{2+}/Ca^{2+}, low E_h and near-neutral pH. These conditions usually restrict siderite formation to non-marine diagenetic environments (low SO_4^{2-}) where abundant Fe^{2+} is present (tropical-zone weathering). The mineral is particularly common in deltaic swamp facies where concretions give evidence for continued growth during progressive burial. For siderite to be stable relative to calcite, the iron concentration must be greater than 5% of that of calcium. In seawater it is less than 0.1%. The occurrence of sideritic beds in certain marine mudstones points to somewhat unusual conditions. Associated burrows and fauna often indicate minor depositional gaps (Sellwood, 1968). The siderite may have formed much later in diagenesis from an oxide precursor when marine SO_4^{2-} was exhausted and when the Ca^{2+} ions were used up.

Chamosite occurs most characteristically as ooids and as mud with siderite in minette-type iron ores. Associated fauna indicate fully marine conditions with agitation to encourage ooid growth. However, the environmental requirements for ferrous silicate stability resemble those of siderite modified by low carbonate activity and saturation with respect to some active silica form. They suggest reducing conditions below the sediment–water interface. It is therefore likely that the chamosite ooids were originally formed of some early mineral (or minerals) that was subsequently converted to chamosite during diagenesis. Sorby originally postulated that the ooliths were formerly calcitic (see Kimberley, 1979), but detailed

Part 3
User's Guide to Sedimentological Fluid Dynamics

4   Back to basics: fluid flow in general

So that water might be
elephantine and pinpoint,
what an industry of air,
what transformation of heat.
And water goes off—it skulks
through the cracks in rocks, jemmying them open . . .

Norman MacCaig, 'Water', *Collected Poems*, Chatto and Windus

4.1 Introduction

Sedimentological fluid dynamics is concerned with *how* the motion of two of the three states of matter, liquid and gas, can transport solid mass in the form of sediment across the Earth's surface. Fluid dynamics is simple enough, but for us as sedimentologists there are added complications that arise due to the effects of the transported solids on the properties of the pure fluid phase.

Liquids and gases share the characteristic that they cannot permanently withstand the action of shearing forces. Thus water in a glass will respond immediately to the slightest tilt or bump. Fluid deformation is quite unlike the recoverable deformation of an elastic solid as typified by the Hookean spring; but, as we shall see, some substances may overlap in their behaviour between liquid and solid. A Newtonian fluid may be defined as a substance that deforms immediately and then continuously when acted upon by a shearing force; the original shape cannot be recovered. It follows that stationary fluids have only normal forces acting upon them. Non-Newtonian flows are deviant in the sense that they disobey the normal rules implicit in our definition—many sedimentologically important examples occur, some of which are highlighted in Table 4.1. We shall return to this topic below.

Also highlighted in Fig. 4.1 are the driving forces that cause the various flows. We can distinguish momentum- and buoyancy-driven flows. Momentum-driven flow is due to externally applied gravity or pressure differences, which set up velocity gradients. There is no density difference affecting the fluid(s) in motion, although in Nature, as we shall see below, admixed solids may often impose their own density

gradients. Buoyant flows depend for their motion on a density contrast with a surrounding fluid.

4.2 Material properties of fluids

In detail the physical material properties of fluids of interest in earth-surface flows depend upon their molecular structure and upon the kinetic interactions of individual molecules. Material in this sense means the *bulk* properties associated with the substance in question: we generally ignore the details of the molecular-scale interactions in sedimentological fluid dynamics.

Density

Density (ρ) is mass (m) per unit volume with dimensions ML^{-3}, in SI units kg/m^3. Density has an important role in fluid dynamics, controlling dynamic quantities such as the *vectors* fluid momentum, force (momentum flux), and static and dynamic pressures and the *scalar* effective (immersed) weight of fluids and solids immersed in or mixed with other fluids. This dynamic role arises because all these quantities depend upon the mass per unit volume, ρ, of fluid that is moving as well as the magnitude of the velocity, u. Water is very much more dense than air and thus for a given velocity the momentum (mu) of unit volume of water is far greater. Similarly the density ratio between the two fluids and quartz varies considerably, a feature that has fundamental implications for the buoyant forces exerted by the two fluids on mineral density grains.

Density is a function of both temperature (Fig. 4.2) and pressure in pure fluids. Under earth-surface

Table 4.1 Earth-surface geophysical flows and their driving mechanisms.

Geophysical flow	Driving mechanism	Flow type
Atmospheric gas flows (e.g. planetary and surface winds)	Atmospheric pressure forces driven ultimately by convection and severely affected by Coriolis force	Newtonian
Surface water flows in channels (river and delta channels)	Gravity	Newtonian
Surface sheet flows (e.g. hillslope runoff, some river floods)	Gravity	Newtonian
Aqueous surface waves of translation (e.g. sea surface waves)	Gravity	Newtonian
Ocean surface currents	Wind shear, pressure gradient and Coriolis force	Newtonian
Tidal flows	Gravity	Newtonian
Surface mass sediment flows (solids only) (e.g. grain flows/rock avalanches; rock flows; some powder snow avalanches)	Gravity	Granular behaviour (Bagnoldian)
Surface mass flows of sediment and water (e.g. debris, turbidity and other mass flows)	Gravity acting upon density contrasts caused by presence of suspended sediment	Non-Newtonian to Newtonian
Deep ocean currents	Gravity acting upon density contrasts caused by salinity differences, differential heating or cooling	Newtonian
Lake and ocean vertical mixing (chimneys, overturns)	Seasonal convection due to salinity differences, heating and cooling of surface waters	Newtonian
Subaqeous internal solitary waves (e.g. reflected density and sediment flows)	Gravity	Newtonian
Near-subsurface (vadose/unsaturated zone) flow in pore spaces	Gravity	Newtonian
Deeper subsurface fluid flow in pore spaces and other discontinuities (e.g. burial diagenesis of aqueous solutions, oil, gas)	Gravity causing compaction and chemical reactions and thermal effects causing density contrasts	Newtonian

conditions, water and air masses are obviously most influenced by temperature changes, the generally negative relationship between density and temperature in both fluids being responsible for the important class of thermohaline currents that dominate deep oceanic circulation (Chapter 26). There are, however, important effects due to the anomalous expansion of water below about 4 °C and to the pressure at depth both in deep lakes (like Baikal) and in the oceans, which decreases the temperature of maximum density by about 0.021 °C/bar. Thus, whilst a water mass near 4 °C at the surface is at maximum density, it is less dense than slightly colder waters at depth due to this thermobaric effect.

Since most natural fluids are far from pure, we also need to consider the effects of contaminants like

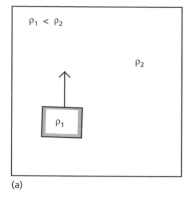

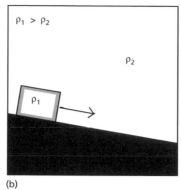

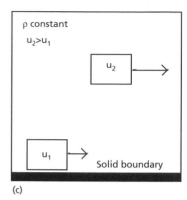

(a)   (b)   (c)

Fig. 4.1 Driving mechanisms for earth-surface flows. (a) Negative buoyancy due to salinity and/or temperature differences. (b) Positive buoyancy and flow down a gravity slope due to temperature and/or salinity differences and/or admixed solids. (c) Viscous retardation (arrow length proportional to velocity, i.e. vectorial) of fluid close to the lower boundary (termed a 'bed' in sedimentological studies or a 'wall' in fluid dynamical literature).

Further information on standard notations

The coordinate axes and vector components that we shall use are shown on Fig. B4.1.

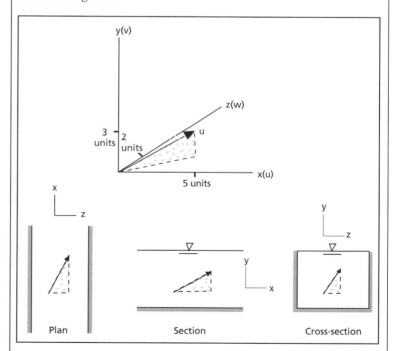

Fig. B4.1 Definition diagram of 3D coordinates (*x, y, z* space) and the corresponding components of the mean velocity vector **u** (*u, v, w*) shown below for a rectangular channel flow. The vector **u** shown has a position in space determined by the ratio of its component parts, 5 : 3 : 2 in this case.

continued on p. 80

1 Space is divided by x, y, z rectangular (Cartesian) coordinates.
2 Local instantaneous fluid velocity is $\boldsymbol{u}$, comprising local instantaneous vectorial components u, v, w corresponding to the x, y, z axes respectively.
3 Local time-mean fluid velocity is $\bar{\boldsymbol{u}}$ about the mean.
4 Any fluctuation in u, v, w (turbulent flows only), with components $\bar{u}$, $\bar{v}, \bar{w}$, is written as u', v', w', with the understanding that the fluctuations may be either instantaneous or are root-mean-square (RMS) deviation samples, i.e. they are time means. To obtain the RMS value, the measured velocity obtained at a point is subtracted from the time-mean value, then squared, then all the squared values are summed and the mean of the squares obtained by dividing by the sample number. The RMS value is the square root of the mean square (see also Chapter 5).
5 The velocity of any solid carried by the flow is U.

Further information on standard notations [*continued*]

dissolved ions, dispersed molecules and transported mineral solids (Figs 4.3–4.5). These effects are absolutely essential in understanding the driving forces behind earth-surface geophysical flows. Concerning the influence of transported solids, we can envisage an effective bulk density (ρ_b) of a control volume. Here, for unit volume,

$$\rho_b = (1 - c)\rho + c\sigma$$

where c is the fractional concentration of solids and σ is the solid density.

Viscosity

Molecular or dynamic viscosity (μ) controls the ease of deformation of a fluid. If this is your first encounter with the concept of viscosity, you can visualize its effects by comparing the force that you would have to exert in stirring substances of different viscosity. You might grunt as you struggle to stir treacle just taken from the refrigerator, but gradually, as the temperature increases (some of the increase is due to

Fig. 4.2 The variation of water and air density with temperature and pressure. (After Vardy, 1990.)

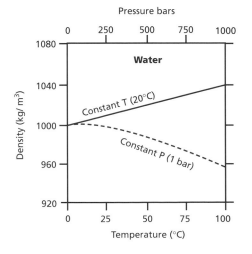

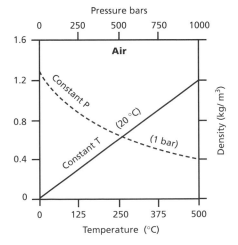

fabric studies do not support this conclusion (Bradshaw *et al.*, 1980). It is possible that the ooids were formed as an Fe-Al-rich gel within wave-agitated lagoons and that the gel was subsequently converted to chamosite during shallow burial (see Curtis & Spears, 1968; Talbot, 1973). However, details of the process remain obscure.

Glauconite formation is poorly understood. It forms as a marine phase in areas of much-reduced sedimentation where it fills in shell cavities and replaces faecal pellets. Berner (1971) notes that it forms slowly at the sediment–water interface where it is associated with organic matter and generally positive but fluctuating E_h conditions. Glauconites vary tremendously in composition from potassium-poor smectites to potassium-rich glauconitic micas with a general trend towards increasing potassium with time.

A major sink for ferric iron during oxygen-poor Archaean and Proterozoic times (*c.* 1.8–3.5 Ga) is recorded by the huge reserves of Fe^{2+} in the banded ironstone formations (BIFs) that are so characteristic of many cratonic nuclei (e.g. Hammersley Range, Australia; Superior Province, North America; Bilengwe Greenstone Belt, Zimbabwe). The iron in these vast deposits is entirely ferric and has usually been thought to have been due to oxidation of ferrous species present in the anoxic depths of the contemporary oceans and brought to the surface in upwelling zones (see Chapter 26 on upwelling). Abiotic oxidation and precipitation in the shallow surface waters of Archaean or Proterozoic oceans then occurred, with the necessary electron transfer being provided either by primitive photosynthesizers or by photochemical (i.e. nonorganic) ferrous iron oxidation. The precipitated species were then deposited below wave base and escaped subsequent reduction during shallow burial.

There are numerous interesting debating points concerning the above ideas, chiefly the contradiction between oxygen-poor atmospheres and massive oxidation required to precipitate ferric iron from dissolved ferrous iron. Most significant is the suspicion that some form of bacterium was actually responsible for the precipitation events, perhaps the purple bacteria discussed previously.

3.14 Phosphates

Phosphates form an important minority of sedimentary rocks. Concentrations of phosphorus in seawater average about 0.07 ppm. Despite these low concentrations, phosphorus is a very important element, being an essential component of all living cells. Increased use of fertilizers and exploitation of natural phosphate reserves have inevitably focused sedimentological attention on the origins of phosphate rock and particularly of the concentration mechanisms involved. Phosphorus turns out to be a key element in tracking the changing magnitude of continental weathering because the weathered flux of phosphorus from the continents dwarfs any primary inputs from the mid-ocean ridges or from atmospheric deposition. Sinks for phosphorus include organic and skeletal matter, iron and manganese oxyhydroxides and the phosphorus-mineral phases francolite–apatite and collophane. Data on phosphorus abundance in ocean-floor sediments show characteristic peaks and troughs (Calovian–Oxfordian, Valanginian, late Albian, Palaeocene), possibly related in a positive manner to larger-scale cycles of sea-level change, i.e. sea level up, phosphorus up and vice versa (Fölmi, 1995). The steady increase of phosphorus accumulation rates since the initiation of 'icehouse Earth' at about 30 Ma occurred during a time of global falling sea levels, perhaps because of the increased efficiency of chemical continental weathering since that time (see also Section 15.8). Phosphate occurrences as phosphorite rocks are either very early diagenetic or replacements after existing carbonate phases. They occur at the sites of present or past oceanic/shelf upwelling and mark periods of great phosphorus enrichment by mass mortality of phosphorus-bearing planktonic organisms.

Further reading

Krauskopf (1979) is still an excellent starting point for elementary aqueous geochemistry and, as in Chapter 2, relevant sections of Atkins (1992), Gill (1989) and Andrews *et al.* (1996) are warmly recommended. Bathurst (1975), though long in tooth, is still delightful. Chapters 11 and 12 in Summerhayes and Thorpe (1996) are excellent on global oceanic geochemistry (Burton) and the carbonate system (Varney) respectively. Tucker and Wright (1990) is an encyclopaedic (but none the worse for that) treatment of all things carbonate.

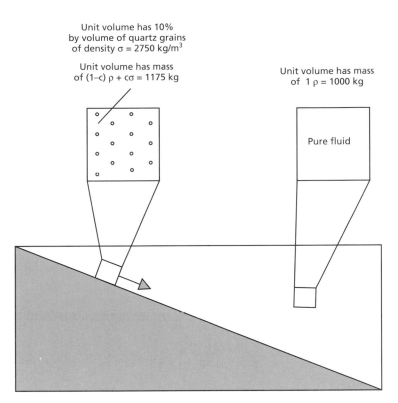

Unit volume has 10%
by volume of quartz grains
of density σ = 2750 kg/m³

Unit volume has mass
of (1–c) ρ + cσ = 1175 kg

Unit volume has mass
of 1 ρ = 1000 kg

Pure fluid

Fig. 4.3 The excess mass of unit fluid volume produced by the addition of transporting solids. The solid–fluid mixture is shown moving down (arrow) the gravity slope due to its excess (positive) buoyancy.

the energy you have used in shearing the fluid being turned to heat), the stirring becomes easier and requires less work. By comparison, how easy it is to stir water (but only with a small spoon) or even easier to stir the air! In more formal words we may say that viscosity controls the rate of deformation by an applied shearing stress, since to set up and to maintain relative motion between fluid layers or between fluid and solid layers requires work to be done against the forces of resistance. The dimensions of molecular viscosity are $ML^{-1}T^{-1}$, in SI units kg/m s or N s/m² (= Pa s). The role of viscosity is best illustrated by Newton's relationship, which we shall come across again later in this chapter:

$$\tau = \mu \frac{du}{dy} \qquad (4.1)$$

where τ is the shearing stress (i.e. the force that you apply whilst stirring), μ is the molecular viscosity and du/dy is the velocity gradient (or strain rate), the relative movement between layers that are unit distance

apart. Molecular viscosity is thus the proportionality factor that links shear stress to the rate of strain. This concept is illustrated in Fig. 4.6. Note that in a fluid it is the rate of deformation, not the actual deformation, that provides the criterion for the stress equilibrium (cf. the modulus of solid rigidity). A succinct (but clinical) definition of viscosity would be: 'the force needed to maintain unit velocity difference between unit areas unit distance apart'.

The causes of viscous forces are complex. In gases with a positive velocity gradient flowing past a surface, molecules continuously diffuse across the flow and effect a net transfer of drift momentum from fast- to slow-moving layers and ultimately to the solid surface. By Newton's second law the rate of destruction of momentum is a force, in this case called the viscous drag. In liquids, in addition to molecular diffusion, there are substantial cohesive attractive forces due to the phenomenon of hydrogen bonding discussed in Chapter 2.

Of the two most important fluids in sedimentology, water is very much more viscous than air, a difference

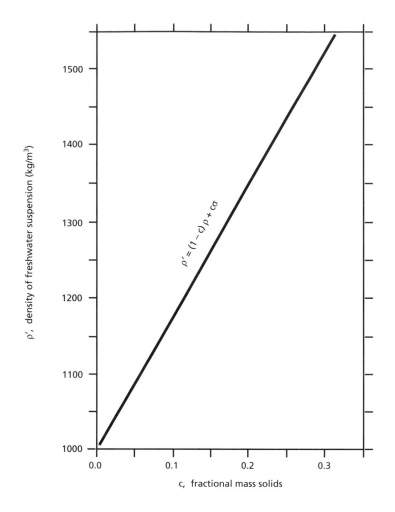

$$\rho' = (1 - c)\rho + c\sigma$$

Fig. 4.4 The variation of freshwater density with volume fraction of added solids of quartz density (2750 kg/m³).

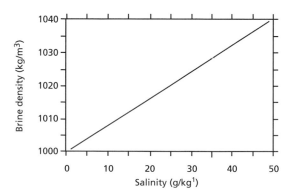

Fig. 4.5 The variation of brine (seawater) density with salinity. (After Denny, 1993).

that is highly important in sedimentological fluid dynamics. Viscosity is a sensitive function of temperature (Fig. 4.7) and the temperature conditions of a particular flow system should always be quoted. In gases, viscosity increases with temperature because more kinetic diffusion occurs. By way of contrast, water viscosity decreases with increasing temperature because the cohesive molecular forces due to hydrogen bonding decrease.

The reader should be aware that another expression of viscosity is in common use. This is the kinematic viscosity, ν (Fig. 4.8), derived for any fluid by dividing molecular viscosity by density, i.e. $\nu = \mu/\rho$. Its dimensions are L^2T^{-1}, in SI units m²/s. The

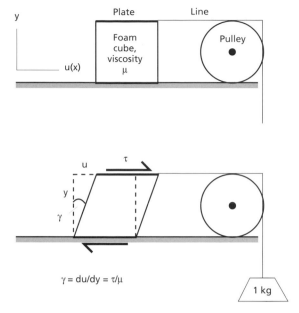

Fig. 4.6 Analogue model for the action of fluid viscosity in resisting an applied force. The force is exerted on the top unit area of the foam cube. In continuous fluid deformation, as distinct from the equilibrium displacement shown here, the displacement in x is the velocity u (as shown).

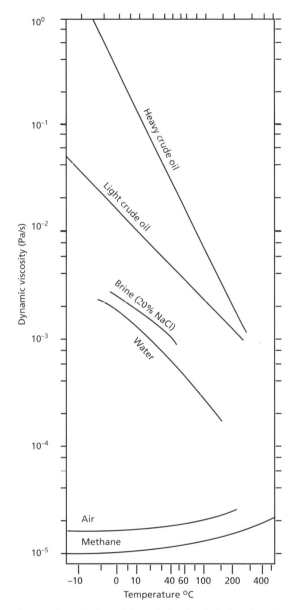

Fig. 4.7 The variation of dynamic (molecular) viscosity with temperature for some natural fluids.

quantity is kinematic because only the units of length and time occur in it; a dynamic quantity must include mass. The term is a useful one because it expresses the ratio between a fluid's ability to resist deformation with a measure of its resistance to acceleration.

As with density, viscosity is strongly controlled by contaminants of various kinds. Most important is the effect of admixed solids. Figure 4.9 shows various estimates of the effect. Perhaps the most famous result is the Einstein–Roscoe equation:

$$\mu_m = (1 - 3.5c) - 2.5\mu \qquad (4.2)$$

where c is the fractional concentration by volume and μ_m is the apparent viscosity of the fluid–solid mixture. Something of the theory behind this and other relationships will be found in Chapter 6, but the basic point is that each solid surface in a fluid is a potential slip plane that increases the internal resistance to shear. So, in order to keep the same velocity gradient overall it is necessary to increase the applied shearing stress. In the homely language of our previous thought experiment, we must stir treacle mixed with peanuts more energetically than with the treacle alone.

Major problems arise in the analysis and characterization of viscosity in those deviant non-Newtonian

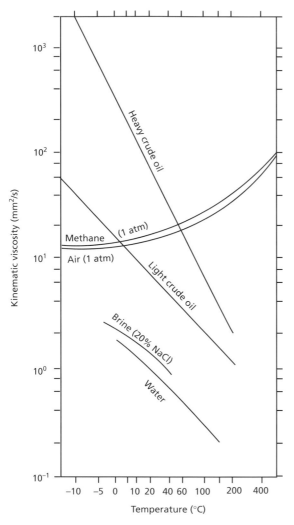

Fig. 4.8 The variation of kinematic viscosity with temperature for some natural fluids.

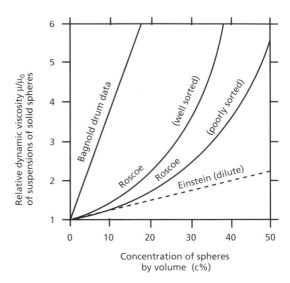

Fig. 4.9 The variation of dynamic viscosity with solids concentration according to a number of theoretical models. The Roscoe curves are most suited to concentrations observed in natural flows. μ is the viscosity of pure water, μ_0 is the viscosity of the mixture.

4.3 Plastic behaviour

As noted previously, Newtonian fluids have a constant viscosity at constant temperature and pressure and suffer continuous deformation (irrecoverable strain) as long as shear is maintained. Plastic substances, on the other hand, show an initial resistance to shear, termed a yield stress, followed by subsequent deformation (Fig. 4.10). Bingham plastics show a constant Bingham viscosity whereas non-Bingham plastics show variable viscosity according to shear rate. It is frequently stated that the addition of solid contaminants to Newtonian fluids may induce a plastic behaviour, although this is certainly not always the case. Plastic behaviour of 'natural' fluids like lava and debris flows has many implications. A finite yield strength allows morphological features like levees, flow snouts and flow wrinkles to be preserved during flow and after motion has ceased. In such flows, particle settling is likely to be hindered or even impossible. If the rate of viscous shear varies across a Bingham flow, as it often does due to the presence of boundary layers (see Chapter 5), then some parts of a shearing Bingham material may be suffering strain

fluids where the shear rate *controls* the viscosity. Names are given to the several distinctive forms of behaviour summarized in Fig. 4.10. Pseudoplastic substances are those where μ decreases as rate of shear increases. Dilatant substances are those where μ increases as rate of shear increases. Thixotropic substances are those where μ decreases with the time for which shearing forces are applied. Finally, in rheopectic substances, μ increases with time as shearing forces are applied.

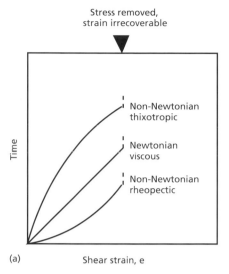

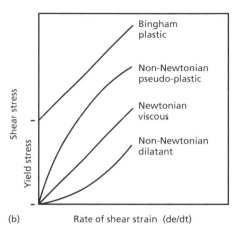

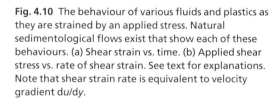

Fig. 4.10 The behaviour of various fluids and plastics as they are strained by an applied stress. Natural sedimentological flows exist that show each of these behaviours. (a) Shear strain vs. time. (b) Applied shear stress vs. rate of shear strain. See text for explanations. Note that shear strain rate is equivalent to velocity gradient du/dy.

whilst others may not. The latter form solid central plugs of material.

4.4 Dimensionless numbers

These are highly important in providing scale- and unit-independent measures of material dynamic behaviour. The Buckingham Π theorems of mechanics state that:

1 All natural phenomena can be described by dimensionless ratios (Π groups).

2 The number of physical quantities exceeds the number of independent dimensionless ratios by the number of fundamental dimensions.

From what we have presented in Table 4.1, relevant Π groups at the Earth's surface will involve the following forces:

• Buoyancy—density-controlled (via temperature, solid/ solute content).

• Viscous—viscosity-controlled (via temperature, solids/ solutes/gas content).

• Inertial—momentum-controlled (via a characteristic velocity and density).

• Gravity—obvious, but note the concept of 'reduced' gravity, whereby a denser fluid moving within a less dense fluid has a gravitational 'pull' that is proportional to the immersed density difference.

To illustrate the point, the ratio of inertial to viscous forces defines the Reynolds number and the ratio of inertial to gravity forces defines the Froude number, both widely used in sedimentological fluid dynamics (for the Reynolds number see Chapter 5).

4.5 Reference frames for flows

The notion of relative motion is a familiar one to the occupant of a stationary vehicle when a neighbouring vehicle moves forwards or backwards. We are initially confused as to just who is moving and it is usually only the absence of an inertial feeling (acceleration or deceleration) that can convince us that *we* are indeed stationary. Now imagine that you are a sedimentologist moving with a sand grain (sitting on it perhaps like a child's bouncing ball) or riding an advancing turbidity current like a surfer. You would 'see' entirely different patterns of flow in the air or water as they pass by you than those observed as the fluid passes by a stationary colleague. This excursion

Further information on Froude number

The Froude number is applicable to water flows having a free surface or interface such that gravity forces play an important role in causing flow. The Froude number is given by the ratio of inertial to gravity forces as given by:

$$\frac{u}{\sqrt{hg}} \qquad (B4.1)$$

where h is a characteristic length, like mean flow depth, g is acceleration due to gravity and u is mean flow speed. The ratio is named after William Froude (pronounced *frood*), a pioneer naval architect who first introduced it. It is an important quantity to consider when scaling down natural flow systems to laboratory models, through the principle that Froude number scaling requires model and prototype to have the same value. The best way of appreciating the significance of the Froude number is to recognize that the expression in the denominator is the velocity of a small surface wave in still shallow water (see Chapter 9). The Froude number is thus the ratio of flow velocity to the velocity of a small wave created in the flow. When the Froude number is less than unity, then the wave velocity is greater than the flow velocity, i.e. waves from a pebble thrown into a flow can travel upstream. Such a flow is said to be tranquil. For a Froude number greater than unity, the flow is said to be rapid. Densimetric Froude numbers may be defined for density currents as:

$$\frac{u}{\sqrt{hg\Delta\rho/\rho}} \qquad (B4.2)$$

where $\Delta\rho$ is the density difference between the different layers of fluid and ρ is the mean density.

into relativity may initially seem trite, but study of Fig. 4.11 will hopefully soon convince the reader that, although the dynamics of the two systems illustrated are similar, the kinematics are not.

Systems where the coordinates x, y and z are fixed in space for a stationary observer are known as Eulerian (pronounced *oilerian*) systems. All kinematic and dynamic analysis is thus done with respect to a fixed control volume through which the fluid passes. Velocity measurements at different times are thus gained from different fluid 'particles'. Systems where the reference axes move with the control volume and are seen by a similarly moving observer are known as Lagrangian systems. All kinematic and dynamic analysis is thus done with respect to a given, constant mass of fluid.

Most systems benefit from an Eulerian treatment, when the velocity at a point is a function of x, y, z and time. The mathematics is easier for this case and we usually wish to consider dynamical results 'at a point', rather than the fate of a single fluid mass. However, it must be stressed that sedimentologists cannot ignore the Lagrangian velocity field, as we shall illustrate in the next section.

4.6 The concepts of flow steadiness and uniformity

Forces, manifested as pressure gradients, are set up in flows that change velocity in time and/or space (Newton's second law). These forces are those which control the magnitude of sediment erosion or

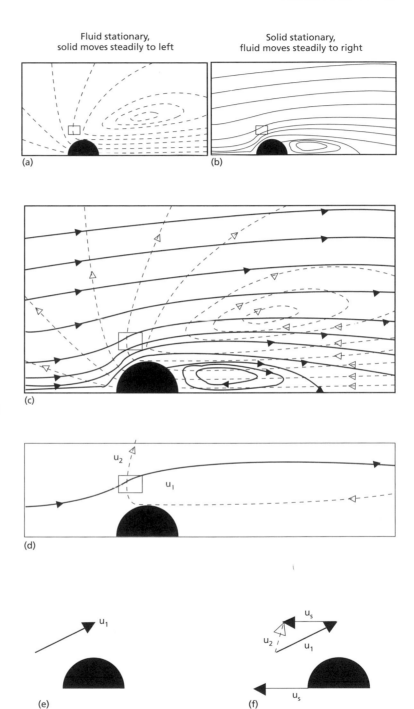

Fluid stationary,
solid moves steadily to left

Solid stationary,
fluid moves steadily to right

(a)

(b)

(c)

(d)

(e)

(f)

Fig. 4.11 Sketches to illustrate relative motion of fluid past a solid hemisphere. (After Tritton, 1988, and sources cited therein.) (a) Case where the fluid is stationary and the solid moves steadily to the left. The flow pattern is that 'seen' by an observer moving with the solid in the Lagrangian fashion. (b) Case where the solid is stationary and the fluid moves steadily to the right. The flow pattern is that 'seen' by a stationary observer in the Eulerian fashion. (c) Superposition of the two flow patterns (a) and (b). (d) Two streamlines selected for analysis of relative motions involved. (e) The velocity vector u_1 for case (b) at the vector tail in the box outlined in (d). (f) To show that when the velocity vector of the moving solid u_s is added to u_1 we find the resultant, which is the vector u_2 appropriate to case (a). This is a general solution applicable to all such cases.

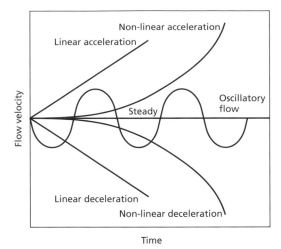

Fig. 4.12 Graphoid (arbitrary coordinate scales) of velocity (measured 'at a point', i.e. Eulerian fashion) vs. time, to illustrate steady, accelerating unsteady, decelerating unsteady and oscillatory flows.

deposition, and thus it is essential that we have a full understanding of the concepts of steadiness and uniformity. Now, a small fluid reference volume may be accelerated or decelerated in two ways.

1 A change of velocity may occur at a given point. Consider yourself a stationary observer who is continuously measuring the velocity, u, of a flow at a point. Should the velocity remain unchanged over the arbitrary measuring interval (within the recording ability of the instrument used), the flow is said to be *steady*, with zero rate of change of velocity with time, i.e. $du/dt = 0$. Conversely, when $du/dt \neq 0$ the flow will be accelerating or decelerating and is said to be unsteady (Fig. 4.12). Thus steadiness is always measured with reference to Eulerian coordinates fixed in the flow boundary at the observation point. Almost all natural flows are, strictly speaking, unsteady, but this clearly depends upon the time interval over which the flow is being measured. Thus tidal flows are unsteady over periods of minutes or hours whereas most rivers are unsteady over periods in excess of days. Turbulence (Chapter 5) leads to unsteadiness over periods of seconds, but we can speak of a steady turbulent flow if the *mean* velocity remains unchanged over the specified time interval.

2 If, by use of several flow monitoring instruments, the observer can make numerous measurements at different places upstream or downstream in a flow and they all indicate the same velocity, then the flow is said to be *uniform* in space, i.e. $du/ds = 0$ (where s is distance). A change of velocity may occur in space at some point downstream or upstream when $du/ds \neq 0$. This is called nonuniform flow and cannot be seen by the stationary Eulerian observer. It is always measured with reference to Lagrangian coordinates, i.e. you must always imagine yourself as travelling with the flow, or having a colleague taking measurements upstream or downstream. A difficulty may seem to arise in boundary-layer flows where viscous effects cause a velocity decrease towards the boundary. Although such flows are clearly nonuniform with respect to the yz plane (normal to the flow direction), it is usual to consider uniformity with respect to the x-axis (downstream). Nonuniform effects thus mostly arise where constrictions or expansions occur in channels, such as where, in the latter case, delta or submarine channels widen downstream before passing into bodies of water of wide extent compared to the channel (Fig. 4.13).

Understanding the concepts of steadiness and uniformity is absolutely vital if we are to understand correctly flow dynamics and the transport, erosion and deposition of sediment. This is simply because of the role of velocity change (accelerations and decelerations) in determining forces exerted by flows. Thus in the most general case the following equation gives the total velocity change at a point, where a is acceleration and s is distance:

$$a = \frac{Du}{Dt} = \frac{u\,du}{ds} + \frac{du}{dt} \tag{4.3}$$

The Du/Dt term is called the substantive (or total) differential. The first term on the RHS of the equation is the advective acceleration term appropriate to nonuniform flows, and records any velocity changes on the way to the observer. The second term is that pertaining at a point to unsteady flows. For uniform, steady flow $a = 0$.

Why should all this be of any interest to the sedimentologist? Well, deposition or erosion at any point of interest results from changes in flow conditions, but are these advective or point changes? The reader will hopefully appreciate that the difference is not a detail of arcane mathematics but is of absolutely fundamental importance! We shall return to the topic

**Further information on
conservation of mass, the general
continuity equation and
divergence** [*continued*]

a sedimentological flow in which density can vary with time due to temperature, salinity or sediment concentration, we can express this condition in symbols as the general continuity equation:

$$\frac{\partial \rho}{\partial t} + \frac{\partial \rho u}{\partial s} = 0 \qquad \text{(B4.3)}$$

Here we have written the velocity with respect to any natural coordinate s. For the simpler case of a steady flow of constant density and resolving the flow components into three dimensions (see Fig. B4.2):

$$\text{div } \boldsymbol{u} = \boldsymbol{\nabla} \cdot \boldsymbol{u} = \frac{\partial u}{\partial x} + \frac{\partial v}{\partial y} + \frac{\partial w}{\partial z} = 0 \qquad \text{(B4.4)}$$

where the inverted delta symbol stands for the divergence. In the 2D flow field illustrated in Fig. B4.2 (could be the shaded portions of Fig. 4.13 or any section through any streamtube), with $u(x)$ and $v(y)$, a flow of velocity $\boldsymbol{u}$ passes in and $\boldsymbol{u} + \delta \boldsymbol{u}$ passes out. Since continuity states that in = out (assuming there are no sources or sinks within the control area), we have through the steps shown the condition div $\boldsymbol{u} = 0$.

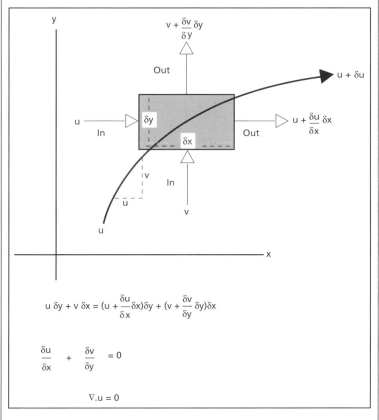

Fig. B4.2 Definition diagram for derivation of the continuity/divergence equation $\boldsymbol{\nabla} \cdot \boldsymbol{u} = 0$.

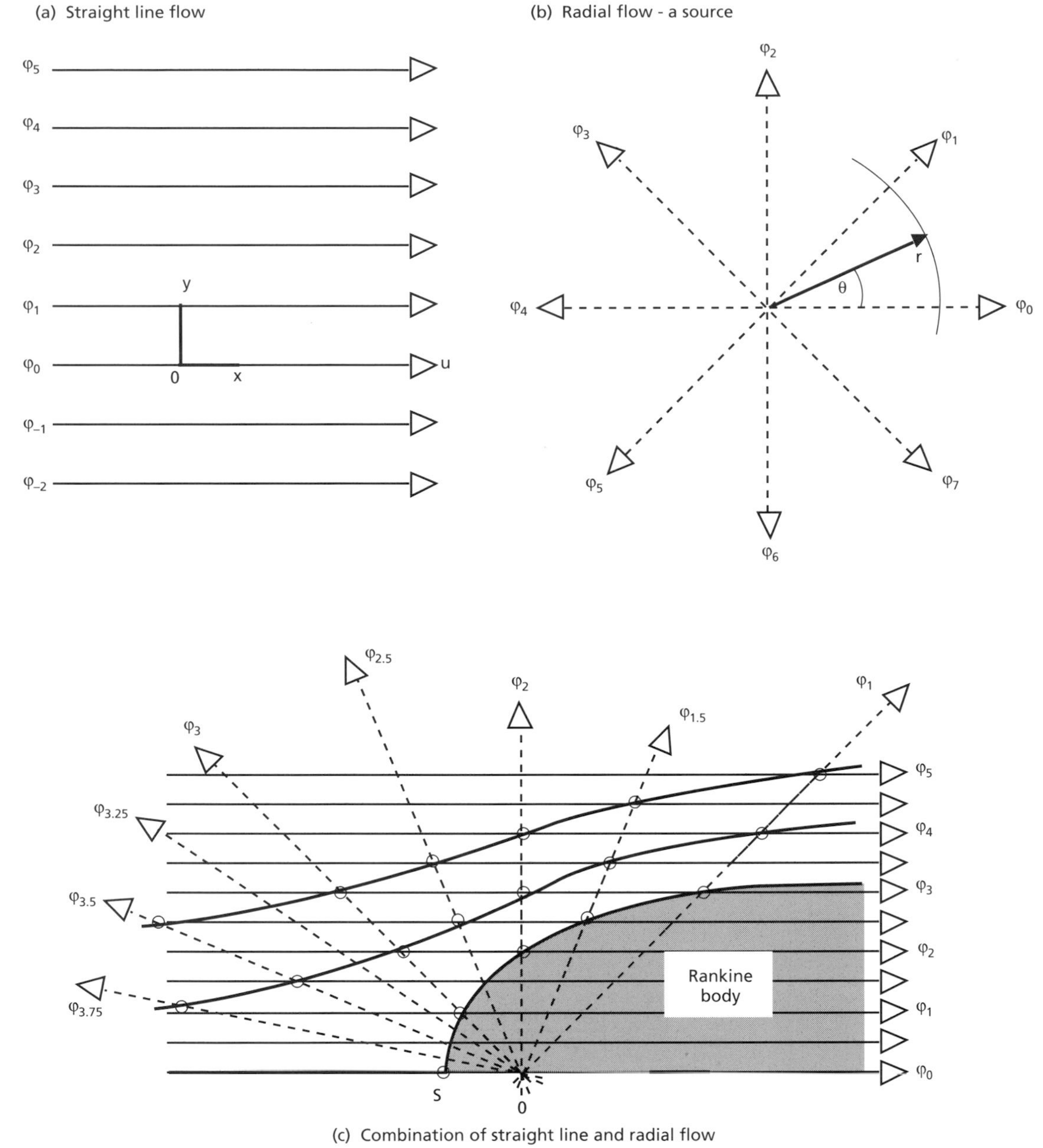

(a) Straight line flow

(b) Radial flow - a source

(c) Combination of straight line and radial flow

Fig. 4.15 Constructed patterns of potential flow for straight line, radial source and combined straight/radial as a Rankine body or, in a sedimentological context, the mean flow streamlines of water over the head of a density current seen in the Eulerian fashion (shaded area).

Further information on vorticity, rotation and circulation

Let us emphasize again that the previous discussion of streamlines, streamfunctions and potentials is applicable only if the fluid is ideal (no shear stresses act on the fluid) and therefore irrotational. When shear stresses act or when rotation occurs, then the streamlines and potential lines are no longer normal. Thus we cannot apply ideal flow very easily to boundary layers. Rotation is best considered with reference to a 2D square element of fluid that is translated, rotated and deformed (Fig. B4.3).

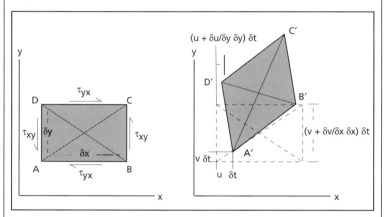

Fig. B4.3 Definition sketch for the mathematical development of vorticity and rotation. (After Vardy, 1990.)

The average rotation (angular velocity $= \omega$) in time is (with anticlockwise positive)

$$\omega = 0.5\frac{\alpha + \beta}{\mathrm{d}t} \tag{B4.5}$$

where

$$\alpha = \frac{\partial v}{\partial x}\,\mathrm{d}t \quad \text{and} \quad \beta = -\frac{\partial u}{\partial y}\,\mathrm{d}t \tag{B4.6}$$

and therefore

$$\omega = 0.5\left(\frac{\partial v}{\partial x} - \frac{\partial u}{\partial y}\right) = 0.5\zeta \tag{B4.7}$$

where ζ is known as the vorticity. In the absence of rotation ζ is zero. Irrotational flow is that in which motion is purely translational, any distortion is symmetrical and vorticity is zero. More generally, for 3D, $\omega = 0.5(\boldsymbol{\nabla} \times \boldsymbol{u})$ and when $\boldsymbol{\nabla} \times \boldsymbol{u} = 0$ (i.e. curl $\boldsymbol{u} = 0$) everywhere, the velocity field is irrotational. It is important to emphasize that rotation defined by vorticity implies a change in spatial orientation, not to circular motion alone. The distinction is well illustrated by the Ferris wheel sketch, whose individual seats undergo no rotation during their revolution. Problems would clearly arise if the seats began to rotate!

continued on p. 96

Rotation occurs in simple shear flow. The action of viscosity will cause simple shear to deform an initial square into a parallelogram in a boundary layer. Here, $\omega = -\partial u/\partial y$ causing one lot of previously normal sides to rotate (Fig. B4.4). Note that a similar shape may be produced by nonrotational symmetrical distortion but that rotation is required to bring the shape into the same orientation as that produced by simple shear. It is sometimes easier to represent rotation with respect to the peripheral velocity of the outside surface of a fluid element. This circulation is the sum of the products of velocity and distance around the outside surface of the element. For 2D rotation about the z-axis, the circulation, G, about an element of area dA is given by Stokes' theorem as $\Gamma = \zeta dA$. In 3D this is usually written as a line integral around a closed path, $\oint \boldsymbol{u} \cdot dr$. In all irrotational flows the circulation is zero.

Further information on vorticity, rotation and circulation
[*continued*]

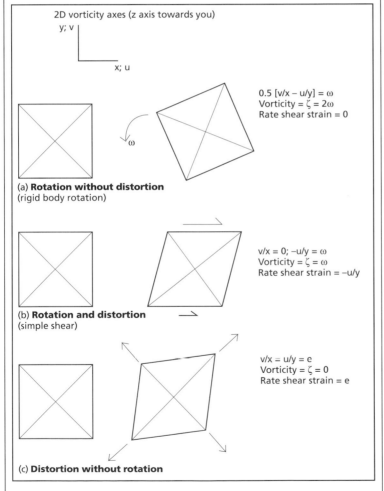

2D vorticity axes (z axis towards you)

y; v

x; u

$0.5 [v/x - u/y] = \omega$
Vorticity $= \zeta = 2\omega$
Rate shear strain $= 0$

(a) **Rotation without distortion**
(rigid body rotation)

$v/x = 0;\ -u/y = \omega$
Vorticity $= \zeta = \omega$
Rate shear strain $= -u/y$

(b) **Rotation and distortion**
(simple shear)

$v/x = u/y = e$
Vorticity $= \zeta = 0$
Rate shear strain $= e$

(c) **Distortion without rotation**

Fig. B4.4 The distinction between rotation with and without vorticity. (After Tritton, 1988.)

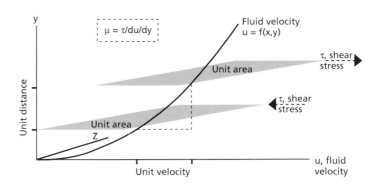

Fig. 4.16 Sketch to show definition of shear stress as a gradient of force acting on unit areas, unit distance apart.

are only a rather simple imitation of natural flow patterns. Experimental studies will reveal the true patterns of flow, in particular the phenomenon of flow separation.

4.9 Dynamics of fluid motion

All fluids in motion, or grains moving in, or being moved by, fluid flow, obey Newton's laws of motion. Every single theoretical result in fluid dynamics, as in all other branches of classical physics, depends upon the application of these laws to particular dynamical problems. Natural flows set up forces due to their changing motion or density. Ample scope exists in Nature for accelerations and decelerations, with density varying according to temperature, presence of dissolved salts, suspended or entrained solid matter. A question often raised is, 'How can a stress be exerted in an ideal fluid when there is no viscosity assumed and no net force acting, i.e. the flow is neither accelerating nor decelerating?' Well, strictly speaking, there is no net force acting because ideal fluids are assumed to be of zero viscosity and to have no friction with their boundaries. In such cases of flow along a solid bed, Newton's third law applies and the fluid force is acted upon by an equal and opposite reactive force from the bed.

But here we must come clean. The movement of fluid past a bed (the basic situation of all sedimentological flows) is seriously affected by friction in a finite zone known as a boundary layer and where the fluid is substantially slowed down with respect to fluid further away from the boundary. We shall look at boundary layers in Chapter 5, but given their general form, with velocity increasing away from the bed, the simplest expression for the force per unit area (i.e. stress, τ) set up in imaginary parallel planes within the flow or the stress exerted by the flow on a solid boundary such as a bed of sand or mud is given by a simple and famous result due to Newton:

$$\tau = \mu \frac{\mathrm{d}u}{\mathrm{d}y} \tag{4.7}$$

where μ is the molecular viscosity. Figure 4.16 reveals the answer as to why net forces are set up in boundary layers: where a velocity gradient occurs, net forces act across shear planes within the fluid. These forces depend upon velocity gradients in time ($\mathrm{d}u/\mathrm{d}t$) *and* space ($\mathrm{d}u/\mathrm{d}y$) and the material property of viscosity.

4.10 Strategies for coping with the dynamic equations

The reader will be reassured to know that we can make a lot of progress by various simplifications of the Navier–Stokes equation. Ignoring viscosity enables us to develop the ideal (potential) flow approach touched on previously. The approach was first proposed by Euler, hence the equation is christened the Euler equation. We write it just for the x-direction:

$$\rho \frac{\mathrm{D}u}{\mathrm{D}t} = \rho g_x - \frac{\partial p}{\partial x} \tag{4.8}$$

We can solve this using simple maths if we assume steady flow between (not across) parallel (not curved) streamlines and replace the x-coordinate by the natural coordinate s, to which we refer each term. The fluid weight term (first term on the RHS) is now written with respect to the local deviation from vertical as

(This is a difficult section, worth persevering with, but one you can ignore if you do not want to get to the very root of fluid dynamics.)

We start with Newton's second law, which leads on to all applications:

$$F = ma \tag{B4.8}$$

The acceleration term

We have already seen how there may be unsteady *and* nonuniform contributions to the acceleration of a fluid element. Remember that a fluid mass may gain or lose velocity either by moving into an area of the flow where the velocity is different or by a change of velocity in time at any position. So we write the total acceleration as Du/Dt, the capital D indicating the total or substantive differential, with s being a natural coordinate:

$$\frac{Du}{Dt} = \frac{\partial u}{\partial t} + \left(\frac{\partial u}{\partial s} \right) \tag{B4.9}$$

This can be expanded into 3D coordinates:

$$\frac{Du}{Dt} = \frac{\partial u}{\partial t} + \left(u\frac{\partial u}{\partial x} + v\frac{\partial u}{\partial y} + w\frac{\partial u}{\partial z} \right) \tag{B4.10}$$

The first term on the RHS is the unsteady part and the bracketed term is the nonuniform part written for a 3D flow. You will often see this written in mathematical shorthand as:

$$\frac{Du}{Dt} = \frac{\partial u}{\partial t} + u \cdot \nabla u \tag{B4.11}$$

We can now write the second law in a new form per unit volume, although no change has occurred to the basic physics:

$$F = \rho\frac{Du}{Dt} \tag{B4.12}$$

The difficulties of the second law for fluid motion arise in part from the nature of the acceleration term, it being a rate of change of velocity both along the flow and at a point. The term may obviously be simplified in very many applications when steady, uniform flow may be assumed, but Nature is rarely so simple!

Let us now look at the innocent-looking F term on the LHS of eqn (B4.8).

The F term

In any nonrotating fluid of constant density, F will comprise forces due to pressure, gravity and viscosity:

$$\sum F = F(\text{viscous}) + F(\text{pressure}) + F(\text{gravity}) = \rho\frac{Du}{Dt} \tag{B4.13}$$

Further information on the equations of motion applied to fluids

continued on p. 99

Further information on the equations of motion applied to fluids [*continued*]

Fluid forces cause a downflow change of pressure, a pressure gradient, to be set up. In the case of a net positive pressure acting on the downstream part of a flow element, we have, per unit volume:

$$F_x(\text{pressure}) = -\frac{\partial p}{\partial x} \tag{B4.14}$$

We have already seen that the stress component parallel to flow in the xy plane due to viscosity acting on unit area of fluid (and directly on any such orientated solid boundary) is given by Newton's equation

$$\tau_{yx}(\text{viscous}) = \mu \frac{\partial u}{\partial y} \tag{B4.15}$$

And the net force acting on any control volume is the difference between the τ_{yx}(viscous) acting on either side of it, i.e. on the rate of change of the viscous stress $\partial \tau_{yx}/\partial y$. For a 2D flow this net force acting in the x-direction is thus

$$F_x(\text{viscous}) = \mu \frac{\partial^2 u}{\partial y^2} \tag{B4.16}$$

per unit volume. Note that the net force is negative for most flows of sedimentological interest because, as we shall discuss further below, the general shape of the velocity distribution in a boundary layer is such that, although velocity increases away from the boundary, the rate of increase $\partial u/\partial y$ gets less. The gravity contribution in the x-direction is given simply as ρg_x per unit volume.

Now we can write the simplest 1D form of the second law for fluids. This is known as the Navier–Stokes equation, written here for a fluid of unit volume, constant density and constant viscosity (i.e. no convection, no temperature changes, Newtonian behaviour, incompressible):

$$\rho \frac{Du}{Dt} = \rho g_x + \mu \frac{\partial^2 u}{\partial y^2} - \frac{\partial p}{\partial x} \tag{B4.\,17}$$

More generally, for 3D flows (rectangular coordinates) and making use of mathematical shorthand:

$$\rho \frac{D\boldsymbol{u}}{Dt} = \rho g + \mu \nabla^2 \boldsymbol{u} - \nabla p \tag{B4.18}$$

the negative slope term ($-dz/ds$). After dividing by ρ and rearranging we then get:

$$u\frac{du}{ds} + \frac{1}{\rho}\frac{dp}{ds} + g\frac{dz}{ds} = 0 \tag{4.9}$$

And after integration with respect to s we get:

$$\frac{u^2}{2} + \frac{p}{\rho} + gz = \text{constant} \tag{4.10}$$

This is now named Bernoulli's equation, after another Swiss mathematician of the 18th century who wrote an early work on fluid flow. It is one of the most important in fluid dynamics, since it relates the pressure and velocity fields in a moving fluid. It is perhaps easiest to appreciate the beauty of the expression by slightly recasting it, multiplying by ρ:

$$\tfrac{1}{2} \rho u^2 + p + \rho gz = \text{constant} \tag{4.11}$$

The three terms on the LHS are now those for kinetic energy, flow energy (best not to use the term 'pressure energy'; see Fig. B4.5) and potential energy respectively. It is easily appreciated that the expression is

The terms 'fluid pressure', 'flow energy' and more especially 'pressure energy' sometimes lead to confusion *if* it is imagined that the pressure in question arises from the ability of the fluid to expand or contract in response to fluid forces. This is *not* the case—most fluids are incompressible and in any case the pressure in question is simply the result of transmitted energy arising from changes in flow pressure, not the energy actually possessed by the fluid. The term 'flow energy' is elegantly explained by reference to Fig. B4.5.

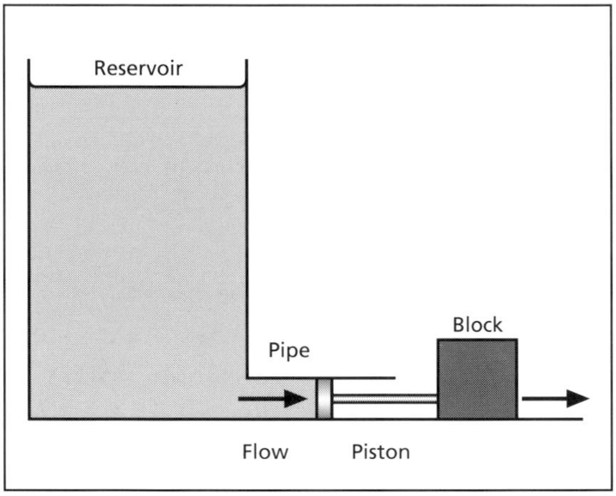

Further information on the energy terms of Bernoulli's equation

Fig. B4.5 How to think of the term 'flow energy'.

simply a statement of energy conservation for a fluid: any changes in one form of energy must be accompanied by changes in the others. However, it is for not just any fluid but one of zero viscosity and constant density, with steady flow along a streamline. The expression should be used with caution in boundary layers with high velocity gradients. A sedimentological example is in order here: if a flow speeds up over a sand grain, then a decrease of pressure results, which may cause the grain to rise from the boundary (see Chapter 6).

introductory texts in fluid dynamics, but both have the needs of the engineer primarily in mind. Denny (1993) has many nice ways of presenting basic fluid physics, but in this case with the biologist in mind. Middleton and Wilcock (1994) is the only text on mechanics (solid and fluid) written with Earth scientists in mind. It is excellent, but requires reasonable levels of and enthusiasm for higher maths. If you are coming to sedimentology from physics or geophysics, then Acheson (1990), Faber (1995) or Tritton (1988) is the book for you. All contain nuggets of clarity.

Further reading

Vardy (1990) and Massey (1979) are clearly written

5 Flow in the real world: laminar and turbulent behaviour

Pride of play in a flourish of eddies,
Bravura of blowballs and silver digressions,
Ringing and glittering she swirls and steadies,
And moulds each ripple with secret suppressions.

Hugh MacDiarmid, 'The Point of Honour', *Complete Poems*, Vol. 1, Carcanet

5.1 Osborne Reynolds and types of flow

In Chapter 4 we made some progress in understanding fluid flow generally, but without real insight into the nature of the flow itself. This was because we simplified the flow conditions so as not to describe the variations of velocity that might occur across a flow (spatial variations) and at any point within the flow (time variations). Major progress on these problems, particularly the latter, came in the late 1870s and early 1880s from a brilliant young English mechanical engineer named Osborne Reynolds. Reynolds set out to investigate the nature of the laws of resistance to flow of fluids in channels and pipes. He approached the problem both theoretically (or 'philosophically' as he put it) and practically, in best physical tradition. His philosophical analysis was '. . . that the general character of the motion of fluids in contact with solid surfaces depends on the relation between a physical constant of the fluid, and the product of the linear dimensions of the space occupied by the fluid, and the velocity'. Designing the apparatus reproduced in Fig. 5.1, he measured the pressure drop over a length of smooth pipe through which water was passed at various speeds (Fig. 5.2). As we have seen in Chapter 4, pressure drop is due to frictional losses as the fluid moves through a system, converting its potential energy to kinetic energy. Reynolds found that the pressure loss per unit pipe length increased with velocity but that at a certain point the losses began to increase more quickly. Below that point, or rather transition region, the trend when plotted on to a graph is of a straight line such that:

$$\Delta p = k_1 u \tag{5.1}$$

where Δp is the pressure loss and k_1 is a constant. Above the transition region:

$$\Delta p = k_2 u^n \tag{5.2}$$

where n is between 1.75 and 2.0 and k_2 is another constant. Deducing that the flowing water was changing its flow pattern, Reynolds confirmed this by introducing a dye streak into a steady flow of water through a transparent tube (Fig. 5.3). At low velocities the dye streak extended down the tube as a straight line, and Reynolds described the flow as 'direct' (now known as *laminar* or viscous flow). With increased velocity the dye streak was dispersed in eddies and eventually coloured the whole flow: this was described as 'sinuous' flow (now known as *turbulent* flow).

The fundamental difference in flow types between the two flow regimes is one of the most important results in the whole field of fluid dynamics. Repetition of the pipe experiments with fluids of different viscosity (the 'physical property' that Reynolds alluded to previously) and different pipe diameters (the pipes always being smooth) showed that the critical velocity for the onset of turbulence was not the same for each experiment. Reynolds found that the change from laminar to turbulent flow occurred at a fixed value of the quantity defined as:

$$\frac{\rho d \bar{u}}{\mu} = \frac{d \bar{u}}{\nu} \tag{5.3}$$

where $\bar{u}$ is the mean flow speed, ρ is the fluid density, μ and ν are the molecular and kinematic viscosity respectively and d is the internal diameter of the pipe. This has become known as the *Reynolds number* in honour of its discoverer. We may think of the Reynolds number as a ratio of two forces acting on

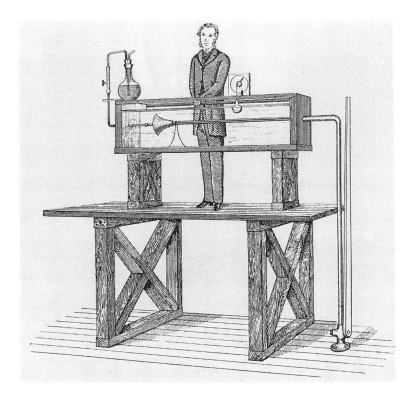

Fig. 5.1 The original sketch of Reynolds' apparatus from 1883.

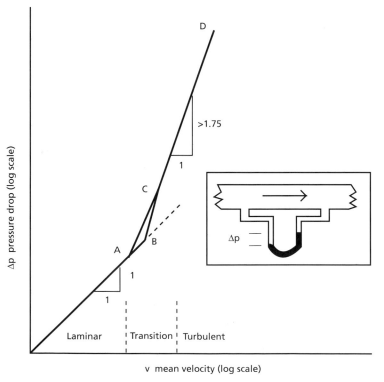

Fig. 5.2 Graphoid to show the flow energy losses (Δp) measured along a smooth pipe (inset) as a function of mean speed (v) through the pipe. (After Massey, 1979.) The rate of loss increases uniformly from the origin to A in laminar flow. The onset and increased rate of loss at ABC (the transition zone) is highly variable, according to local conditions, but eventually at C the rate of loss stabilizes at some higher constant slope in the turbulent state of flow.

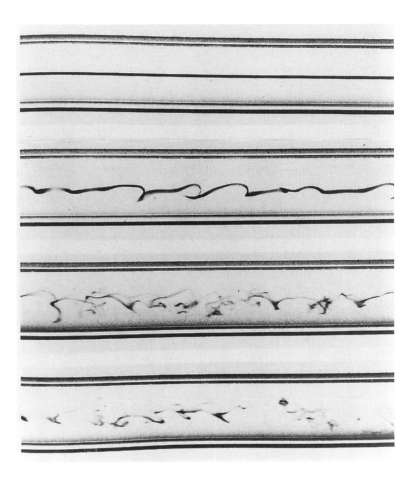

Fig. 5.3 A modern replication of Reynolds' experiment. Top to bottom shows laminar, transitional and turbulent flow patterns visualized by a central dye streak deforming in water flowing left to right through a glass tube. (From van Dyke, 1982.)

the fluid. Viscous forces will resist deformation of the fluid: the greater the molecular viscosity, the greater the resistance. Inertial forces represent the ability of the fluid mass to accelerate. The Reynolds number may be derived from first principles as shown in the box below.

When viscous forces dominate, as say in the flow of liquefied mud or lava, then *Re* is small and the flow is laminar. When inertial forces dominate, as in the atmospheric flow of air and most water flows in rivers, then *Re* will be large and the flow turbulent. For flows in pipes and channels, the critical region for the laminar –turbulent transition lies between 500 and 2000.

We should be careful in any identification of laminar flow with high-viscosity liquids alone. As Reynolds himself took great pains to emphasize, the flow state is dependent upon four parameters of flow, not just one. Thus a very low density or very low velocity of flow has the same reducing effect on

Reynolds number as a very high viscosity. Thus as Shapiro (1961) in his classic introductory text states '. . . it is more meaningful to speak of a very viscous situation than a very viscous fluid'. Flow systems with identical Reynolds numbers are said to be dynamically similar, a feature made use of in many modelling experiments.

Some comment is also necessary on the length scale in the Reynolds number criterion. For a pipe this is the diameter, but for a river channel or other free-surface liquid flow the mean flow depth is an appropriate scale. For wind the length scale presents more problems. In air tunnels this may be taken as the height of the tunnel, but in atmospheric flows the boundary-layer thickness must be chosen. For a Reynolds number applied to a grain of sediment in a flow, the grain Reynolds number, the mean diameter of the grain is the length scale with which to consider flow interactions with the whole grain volume.

This is a simple way of deriving the Reynolds number, following Reynolds' own outline sketch in his classic paper of 1885. We seek to express the balance of inertial to viscous forces acting on a fluid volume. Consider a cubic fluid volume of viscosity μ, density ρ, cross-sectional area δ^2 and length δ as shown in Fig. B5.1. Let the upper surface of the volume move with speed u relative to the lower surface, the gradient of speed being due to viscous forces. Motion may also cause forces to be set up, in this case inertial or due to the fluid's own change in momentum that occurs during acceleration or deceleration.

More information on derivation of Reynolds number

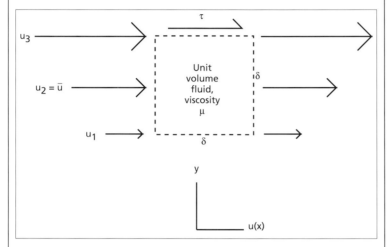

Fig. B5.1 Definition diagram for a derivation of the Reynolds number.

The viscous force acting on the xz plane is equal to the shear stress times area, $\tau\delta^2$. But $\tau = \mu(du/dy)$ by Newton's viscous equation and u/δ is the gradient across the cubic element in the xy plane, so we have:

$$\text{viscous force} = \frac{\mu u \delta^2}{\delta} = \mu\delta u \tag{B5.1}$$

The inertial force is equal to mass × acceleration by Newton's second law. Now mass = density × volume = $\rho\delta^3$ and acceleration = $u/t = u^2/d$ (since $u = \delta/t$ and $t = \delta/u$). Thus:

$$\text{inertial force} = \frac{\rho u^2 \delta^3}{\delta} = \rho\delta^2 u^2 \tag{B5.2}$$

Arranging the two forces as a ratio gives the Reynolds number as:

$$Re = \frac{\rho u^2 \delta^2}{\delta\mu u} = \frac{\rho u \delta}{\mu} \tag{B5.3}$$

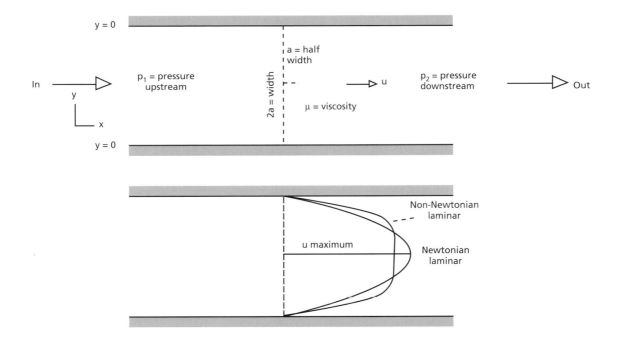

Fig. 5.4 Definition diagram for the derivation of parabolic Newtonian laminar and plug-like non-Newtonian laminar flow.

5.2 The distribution of velocity in viscous flows: the boundary layer

As we briefly noted towards the end of Chapter 4, when real fluids move, frictional effects cause the production of zones of flow retardation adjacent to stationary boundaries of the flow. The zones of retardation are known as *boundary layers*. We are all familiar with boundary layers from everyday experience; thus our inclination to duck down close to the ground when the wind blows exceptionally strongly. The properties of boundary layers have a number of very important consequences in sedimentology because much sediment transport and bedform development occurs in the boundary layer itself.

Physically, the boundary layer is a zone where there is a velocity gradient, i.e. $du/dy \neq 0$, and thus where viscous and inertial forces give rise to shear stresses. Close to the wall or flow boundary we can imagine that there is an attached layer of liquid molecules: the *adsorbed layer*. As the liquid flows, the velocity tends to zero at the wall because the adsorbed layer refuses to move because of surface attraction to the solid atoms (true generally only below a certain critical shear rate). This is the no-slip condition first proposed by Navier and developed notably by Reynolds. This viscous retardation gradually dies out away from the wall until at some point in the flow, termed the free stream, there is no velocity gradient and hence no stress. At least that is the theory. In practice there is never complete retardation, nor is there any zone free from the influence of the boundary. But we make progress by approximations to the truth, and it is the exact rate of change of velocity with distance from the solid boundary that is of interest. If it is a linear change, then the shearing stresses set up by viscosity will everywhere be equal. If on the other hand it is nonlinear, then there will be an important localization of stresses either close to the boundary or far from the boundary depending on the sign of the velocity function.

We need to consider the velocity distribution in a laminar flow between two walls, as in a channel. If we measure point velocities across such a flow, then a characteristic parabolic curve results. This parabolic curve may be exactly predicted by simple theory (another result due to Reynolds) when the pressure and viscous forces acting on the fluid are balanced (Fig. 5.4).

Referring back to the Navier–Stokes equation of Chapter 4, in a steady uniform two-dimensional flow the acceleration term is zero, so the equation written for flow in the x-direction becomes:

$$\mu \frac{\partial^2 u}{\partial y^2} - \frac{\partial p}{\partial x} = 0 \qquad (B5.4)$$

or, with

$$-\frac{\partial p}{\partial x} = -\frac{dp}{dx} = -\frac{p_1 - p_2}{l} = -p$$

we get

$$\frac{\partial^2 u}{\partial y^2} = -\frac{p}{\mu} \qquad (B5.5)$$

We can solve for the distribution of u along the y-axis (across the flow) by integrating:

$$\frac{du}{dy} = \int \left(-p \frac{1}{\mu} \right) dy = -\frac{py}{\mu} + c_1 \qquad (B5.6)$$

and since $du/dy = 0$ at $y = 0$, we find $c_1 = 0$. Integrating again we have:

$$u = \int \left(-\frac{py}{\mu} \right) dy + c_2 = -\frac{py^2}{2\mu} + c_2 \qquad (B5.7)$$

and since $u = 0$ at $y = \pm a$, we find

$$0 = c_2 - \frac{pa^2}{2\mu} \qquad \text{so} \qquad c_2 = \frac{pa^2}{2\mu}$$

and we finally arrive at

$$u = -\frac{p}{2\mu}(y^2 - a^2) \qquad \text{or} \qquad u = \frac{p}{2\mu}(a^2 - y^2) \qquad (B5.8)$$

So at the end of all that we have the result that u is at a maximum for $y = 0$ and $u \rightarrow 0$ as $y \rightarrow \pm a$. The overall form of the velocity profile is a parabola, as the reader may confirm by entering values into the expression. The above derivation is a wonderfully economical generalization from basic physics and is due to Reynolds himself in 1895.

It should be noted that the expression for viscous flow is only strictly applicable to Newtonian fluids at low Reynolds numbers. A case in point would be the *very* slow flow of water along a deep channel. In the boundary layers of non-Newtonian fluids like debris flows the velocity gradients, and hence the strain rates, are much higher at the margins of the flow and much lower in the centre of the flow (Fig. 5.4). In fact, the central zone hardly shows differential motion at all, tending to move *en masse* as a plug. Plug

behaviour in debris flows gives rise to a number of features of sedimentological interest (Chapter 11).

5.3 Turbulent flow

Turbulent flows dominate earth-surface transport processes—most wind and water flows are turbulent. Much of the necessity for understanding turbulence came originally from the field of aeronautics, and it is perhaps no coincidence that 'modern' fluid dynamical

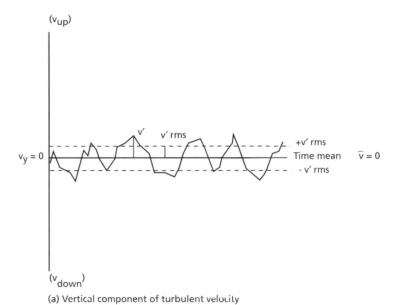

Fig. 5.5 Graphoids to show the various components of turbulent flow. Note the difference between an instantaneous velocity fluctuation, *u'* (positive or negative about the mean), and a measure (root-mean-square) of the mean fluctuation, *u'*$_{rms}$.

(a) Vertical component of turbulent velocity

(b) Horizontal (streamwise) component of turbulent velocity

analysis of turbulence started around the date of *H. sapien*'s first few uncertain attempts at controlled flight. Eighty years later photographs of the effects of turbulent atmospheric flows on Earth were taken by people standing on the Moon, and radar can now reveal Venusian and Martian dunes and sandstorms.

Insertion of a sensitive flow measuring device into a turbulent flow for a period of time will result in a fluctuating record of fluid velocity (Fig. 5.5) regardless of the fact that the flow is steady in the mean. The longitudinal *x* velocity, *u* (sometimes referred to as the streamwise velocity component), is equal to the mean velocity $\bar{u}$ plus or minus the instantaneous deviation *u'* from the mean. In symbols:

$$u = \bar{u} \pm u' \tag{5.4}$$

Thus we can only really talk about a characteristic time-mean flow velocity for turbulent flows, with the mean measured over several tens of seconds or longer depending on the nature and size of the flow. It is important to emphasize that a turbulent flow may very well be steady in the mean, i.e. repeated measurements of $\bar{u}$ are equal. Also, since *u'* may be positive and negative about the mean at different times, we have over a long time period, $\overline{u'} = 0$.

In any turbulent flow (e.g. Fig. 5.3) it is apparent that fluid is moving around in what initially seem to be confusing patterns. Careful analysis shows a well-defined coherence or structure to the motions, a feature we shall consider later (see Section 5.6). At one moment a small volume of fluid is close to a solid boundary, well within the boundary layer, and then it

The following quantities are obtainable from 2D (*xy* plane streamwise/normal to flow bed) measurements of velocity in turbulent flows. The quantities are discussed as they arise in this chapter.

- *u'* and *v'* are velocity fluctuations about a time-mean value, either used as instantaneous values *or* sometimes calculated as root-mean-square fluctuations u'_{rms} and v'_{rms}, i.e.

$$u'_{rms} = \left[\frac{1}{n}\sum(u - \bar{u})^2\right]^{0.5}$$

See individual cases for usages. Note that the mean of all summed (but *not* squared) fluctuations is always zero.
- $-\rho\overline{u'v'}$ is the Reynolds' stress (N/m²) per unit fluid volume, i.e.

$$-\rho\overline{u'v'} = \left[\frac{1}{n}\sum(u - \bar{u})(v - \bar{v})\right]$$

with *u'* and *v'* being instantaneous fluctuations.

Further information about turbulent velocity quantities

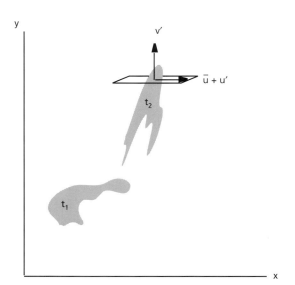

Fig. 5.6 Simple sketch diagram for fluid momentum transfer across unit area of the *xz* plane by an eddy moving in *xy* velocity space from time t_1 to time t_2. It carries time-mean *x* momentum of $-\rho\overline{u'v'}$. (After Massey, 1979.)

suddenly shoots across the flow at high velocity. Such fluid eddies cross the time-mean streamlines that we could visualize. Referring back to Reynolds' original experimental results, these suggest that eddies trans-

fer mass and thus they must also transfer momentum and redistribute it within the flow (Fig. 5.6).

Since we know that rate of transfer of momentum is a force, it is clear that additional turbulent stresses will be set up in turbulent flows, additional that is to those created by dynamic viscosity. These additional turbulent stresses that cause an apparent increase of energy loss were termed the 'eddy viscosity' by their discoverer Boussinesq in 1877.

Turbulent eddies have a 3D form, and in addition to the *x*-component (eqn 5.4) of velocity there will be *y*- and *z*-components as well:

$$v = \bar{v} \pm v' \qquad \text{and} \qquad w = \bar{w} \pm w' \qquad (5.5)$$

Fortunately, *v* and *w* are usually small in comparison with *u* for most shear flows of sedimentological interest. For straight-line flow the mean values of *v* and *w* are zero, i.e. there is no net upward or spanwise transport of fluid. Such 3D turbulent fluctuations cause local velocity gradients to be set up in the flow: these in turn cause local stresses, which work against the mean velocity gradient to remove energy from the flow. The action of these local stresses provides turbulent energy, which is ultimately dissipated by the action of viscosity on the turbulent fluctuations. Thus we now have an explanation of the greatly increased energy losses in Reynolds' pipe flow experiment for

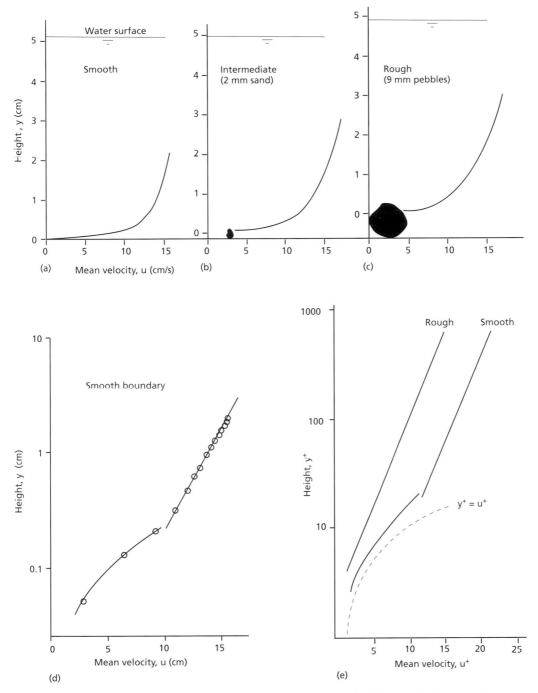

Fig. 5.7 (a)–(c) Graphs to show mean turbulent velocities $\bar{u}$ measured over smooth, intermediate (sand grain for scale) and rough (pebble for scale) boundaries. (d) Smooth boundary data of (a) replotted on semilog ordinate scale to show viscous sublayer and the log-linear distribution in the remaining flow (note deviations increasing towards the surface). (e) Rough boundary data (plotted using dimensionless scales: $y+$ is yu_*/n; $u+$ is $\bar{u}/u_*$) superimposed on curve from (d). Note absence of viscous sublayer. (Data of Grass, 1971.)

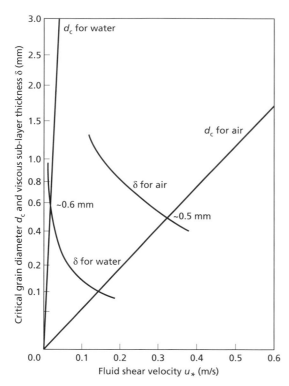

Fig. 5.8 Plot of fluid shear velocity, u_*, against grain diameter, d, and viscous sublayer thickness, δ. Intersections of the δ and d curves for air ($\approx$0.5 mm) and water ($\approx$0.6 mm) define the value of grain size, d_c, at which the grains will cause disruption of a smooth boundary at the threshold of motion. (After Carson, 1971.)

Now, armed with the quantity u_* we can develop the concept of flow boundary roughness by comparing the size of boundary irregularities or sediment grains to the thickness of the viscous sublayer of turbulent flows. First, knowing $u = \tau y/\mu$ (eqn 5.8) we can write:

$$\frac{u}{u_*} = \frac{u_* y}{\nu} \tag{5.12}$$

Perusal of Fig. 5.7 shows that eqn (5.12) closely fits the experimental data up to a value of u/u_* of about 11.5 for a perfectly smooth boundary. Thus we can say that the viscous sublayer in a turbulent flow must have a thickness

$$y = \delta = \frac{11.5\nu}{u_*} \tag{5.13}$$

Figure 5.8 shows how very thin the viscous sublayer actually is. This brings us to an important point. Surfaces whose roughness elements of sedimentary particles are enclosed entirely within the viscous sublayer are said to be *smooth* and are only subject to viscous forces. When the particles project through the sublayer, as assumed in the development above, they shed off eddies whose size increases directly with that of the particles. The surface is then said to be *transitional* or *rough*, depending on the degree of penetration, and the rate of energy loss by turbulent eddy-shedding is higher. The graph of Fig. 5.8 shows d in relation to flow strength for sand grains in air and water flows, plotted so that the critical u_* is that appropriate to the threshold of grain motion (see Chapter 6). In water, for example, the flow boundary ceases to behave smoothly at the threshold of motion for all grain diameters greater than about 0.6 mm; important sedimentological consequences follow from this fact.

Stationary grain roughness of the kind noted above (generally termed *skin friction*) is not the only means by which energy is dissipated in a moving fluid. Sediment transport produces a moving bedload layer that further dissipates energy. Bedforms also add a significant component of friction, known as 'form drag', which arises from the generation of eddies due to the phenomenon of flow separation. Finally, energy is lost whenever a channel flow turns a bend or constricts or expands. All these energy losses may be lumped into one term, a generalized friction coefficient for a particular flow in a particular context. The dimensionless friction term, f, is simply given by the ratio of the shearing stress, or drag, τ_0, exerted by a flow on a surface to the mean kinetic energy per unit volume of the flow, $0.5\rho\bar{u}^2$. That is:

$$\tau_0 = f(0.5\rho\bar{u}^2) \tag{5.14}$$

You may see this general expression referred to as the 'quadratic stress law'. The friction factor, f, takes the form of an experimentally determined drag coefficient, C_D, for objects falling through stationary flows, whilst for channelized flows the constant of proportionality becomes $0.25f$ in the expression:

$$\tau_0 = 0.25f(0.5\rho\bar{u}^2) = 0.125f(\rho\bar{u}^2) \tag{5.15}$$

and is known in hydraulics as the Darcy–Weisbach friction coefficient. The quadratic term in $\bar{u}$ has important repercussions for sediment transport by extreme flows. In general terms, f is strongly dependent on the Reynolds number for laminar to transitional flow regimes. For fully turbulent flows f is more or less independent of Reynolds number, being constant for a particular relative roughness (ratio of roughness diameter to flow thickness) and increasing with relative roughness. (We may note in passing that surface drag 'management' is a practical topic of great importance, form drag variations for different shaped vehicles being a well-known example. Less well known are the dimpling of a golf ball to decrease form drag and obtain greater range, and the natural evolution of certain patterns of roughness elements on fish skin that reduce surface skin friction drag.)

Now consider a steady, uniform, open-channel water flow of average depth h. The downslope shear component, τ_0, of the normal stress, ρgh, due to the water acting on unit area of the bed is simply:

$$\tau_0 = \rho ghS \tag{5.16}$$

where S is the water slope (= bed slope for uniform flow). Equation (5.16) is the du Boys tractive stress equation. The Darcy–Weisbach and du Boys equations can be combined to give the useful expression for mean flow velocity:

$$u = \sqrt{\frac{8g}{f}}\sqrt{Rs} \tag{5.17}$$

where $\sqrt{8g/f}$ is the Chezy coefficient and R is the channel hydraulic radius, usually approximated as the mean depth in wide natural channels.

In everyday language we refer to flows as being fast or slow, strong or weak. In fact, there are a number of alternative parameters that can be used to express flow magnitude in a more exact way. We have already come across mean velocity, $\bar{u}$, and bed shear stress, τ_0. The product of these two parameters gives us the stream power, ω (dimensions MT^{-3}) available to unit bed area of fluid:

$$\omega = \tau_0\bar{u} \tag{5.18}$$

In a channel the available power supply Ω to unit length is the time rate of liberation (in kinetic form) of the liquid's potential energy as it descends the gravity slope S. Thus:

$$\Omega = \rho gQS \tag{5.19}$$

where Q is the whole discharge of the stream. The mean available power supply ω to the fluid column over unit bed area is thus:

$$\omega = \frac{\Omega}{\text{flow width}} = \frac{\rho gQS}{\text{flow width}} = \rho gdS\bar{u} = \tau_0\bar{u} \tag{5.20}$$

The concept of available fluid power is an important one since Bagnold has made extensive use of it in his sediment transport theory (Chapter 6).

5.6 The periodic coherent structures of turbulent shear flows

We have mentioned several times that Reynolds' 'sinuous motion' that we now know as turbulent flow is dominated by eddy motions with vorticity. Prandtl also noted these eddies and from his surface flow visualization experiments regarded them as 'lumps' of fluid (*Flüssigkeitsballen*) transferring momentum through the boundary layer. The eddies gradually develop as the Reynolds number increases from within a stable laminar flow field. We would like to know a little more about the nature of these eddies. For example:

1 Are the eddies random and hence unpredictable in time and/or space, or are they coherent entities, with structure?

2 If there is such a thing as a coherent eddy, what does a 'typical' eddy look like in two or three dimensions?

3 How do eddies shed light on ideas of momentum transfer in turbulent flows and the true nature of turbulent stresses?

4 What relevance have eddies to sediment transport and bedform theory?

Looking at flow visualization results in 2D flows in xy and xz space (Figs 5.9 & 5.10) we may make the following simple observations.

1 In the xy plane:

(a) Fast 'lumps' of fluid move down from the outer to the inner flow region. These are called sweep motions.

(b) Slow 'lumps' move out from the inner to the outer flow. These are called burst motions.

2 In the xz plane:

(a) Close to the bed there exist flow-parallel lanes of slow and fast fluid that alternate across the flow. The low-speed lanes are termed streaks.

(b) These lanes become increasingly less well defined as we ascend the flow.

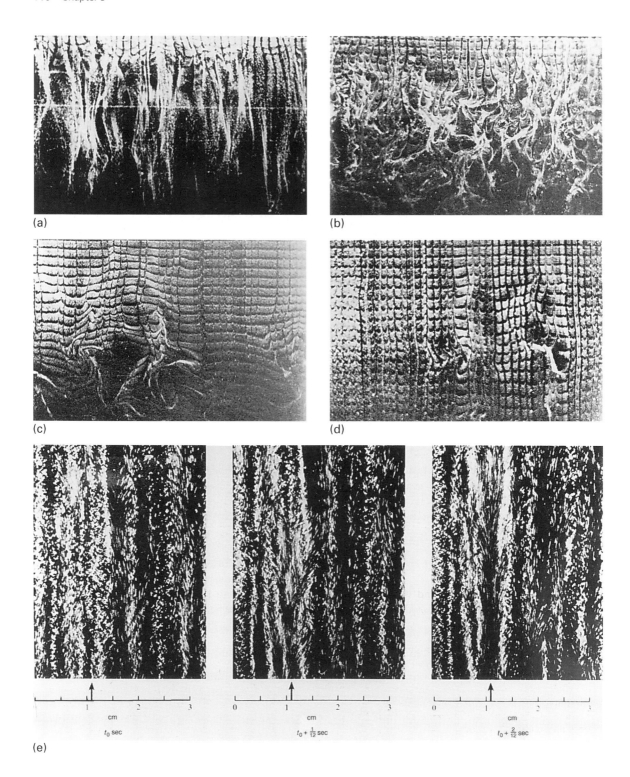

(a)

(b)

(c)

(d)

(e)

Fig. 5.10 Turbulent eddies in air flow (flow left to right) visualized by smoke injected from the bed. (a) An instantaneous view from above to show the streaky inner zone structure (the thin straight white line is the beam of light used to visualize the vertical section in (b)). (b) A simultaneous sideways view showing the beautiful form of periodic ascending low-speed burst elements (white) and descending high-speed outer and transitional layer sweep fluid (black). See the sketch of Fig. 5.11 for a 3D view. (After Falco, 1977.)

A 3D reconstruction of the eddies of turbulent shear flows (Fig. 5.11) shows the presence of large-scale coherent vortex structures within the boundary layer. These are rather far removed from Prandtl's 'lumps' of fluid and more similar to Reynolds' 'sinuous' description. They have the shape of hairpins whose 'legs' are formed from the low-speed streaks. The streak spacing λ_s in the viscous sublayer is given by the experimental relation

$$\frac{\lambda_s u_*}{\nu} = 100 \tag{5.21}$$

where u_* is the shear velocity and ν the kinematic viscosity. Thus with constant ν, λ decreases as u_*

increases. The rate of bursting of low-speed streaks increases with u_*. It should be noted that the streak pattern is quasicyclic, new streaks forming and re-forming constantly across the flow.

We may summarize the above discussion by dividing the turbulent boundary layer into two rather distinct zones: (i) an inner zone close to the bed with its upper boundary between the transition and logarithmic region of the turbulent boundary layer; and (ii) an outer zone extending up to the flow free surface.

1   The inner zone is distinguished by:
   (a) being the site of most turbulence production;
   (b) containing low- and high-speed fluid streaks that alternate across the flow;

Fig. 5.9 (*opposite*) (a)–(d) Instantaneous photographs of H_2-bubble blocks in water, taken from above (looking down towards the bed), the field of view being in the *xz* plane with current flow from top to bottom of each photograph. The speck-insulated platinum wire where the bubble blocks are periodically generated is at the top of each photograph. The sequence (a)–(d) represents successively higher positions of the wire above the bed. In (a) the streaky deformation of the bubble blocks is well shown, each streak identifying a low-speed phase of the cycle in the viscous sublayer. In (b) the streaks have become

tangled and less obvious as they pass into the logarithmic portion of the turbulent boundary layer. In (c) and (d) the blocks are mostly undisturbed in the outer regions of the flow but larger areas of macroturbulence are prominent. (After Kline *et al.*, 1967.) (e) Viscous sublayer structure visualized by means of 0.1 mm diameter sand moving over a smooth black boundary. The sequence of photographs, separated in time by 1/12 s with a 1/30 s exposure, illustrate the development of an inrush or 'sweep' event (arrowed). (After Grass, 1971.)

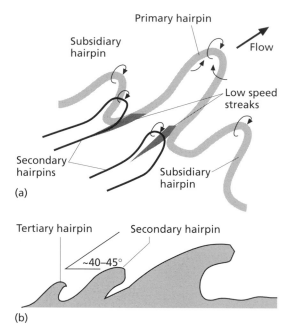

(a)

(b)

Fig. 5.11 (a) Sketch to show 3D configuration of hairpin vortices of turbulence, with the low-speed streaks occurring on their stretched lower limbs. (b) Sketch parallel to flow (*xy* plane) to show lower boundary-layer fluid (shaded) of burst motions (erupting hairpins) and faster downward-moving outer flow fluid of sweeps alternating streamwise along the flow. (After Smith & Walker, 1990; Best, 1993.)

(c) lift-up of low-speed streaks in areas of high local shear near the upper boundary.

2 The outer zone is distinguished by:

(a) providing the source of the high-speed fluid of the sweep phase near its lower boundary that probably initiates a burst cycle;

(b) containing large vortices near the area of burst break-up that are disseminated through the outer zone and may reach the surfaces as 'boils'.

The results all too briefly summarized above are the work of nearly 30 years of experiments into the structure of turbulent flows. Although the mathematical development of hypotheses for turbulent behaviour remains limited, direct observation and analysis of turbulent structure is now possible using sophisticated flow visualization techniques. The results suggest high positive contributions to local Reynolds stresses across the whole flow depth due to burst phases when migration of momentum-deficient fluid occurs. The inrush/sweep phases also give a positive contribution to Reynolds stresses, but their effect is at a maximum in the area close to the wall (Fig. 5.12). Almost all (> 70%) of the Reynolds stresses in turbulent flows are due to burst/sweep processes and the majority of these stress contributions are produced close to the wall. The nature of bursts and sweeps also explains why the bed-parallel Reynolds shearing stresses act in the same direction as the viscous bed shear stress. Remember that in the expression for the total shearing stress, $\tau_{yx} = \mu\, \partial u/\partial y - \rho \overline{u'v'}$, the last term is written as negative because it has been brought over from the LHS of the standard Navier–Stokes expression for acceleration.

The use of a simple quadrant diagram (Fig. 5.13) brings out the essential contrasts between burst and sweep turbulent interactions. Turbulent bursts are *quadrant 2* events because they involve injections of slower-than-mean horizontal velocity (i.e. instantaneous u' values are negative) fluid *upwards* (i.e. instantaneous v' velocities are positive), the average product is negative and thus $-\rho \overline{u'v'}$ is overall positive.

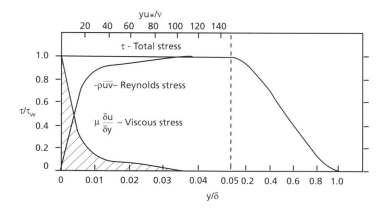

Fig. 5.12 Distributions of the total stress (τ), Reynolds stress ($-\rho \overline{u'v'}$) and viscous stress $\mu(\partial u/\partial y)$ across a turbulent boundary layer (Reynolds number of 7×10^4). Note 30-fold change in the abcissa scale at $y/\delta = 0.05$. The only stress exerted directly at the smooth bed is a viscous one, but just away from the bed turbulence generates an increasingly large Reynolds stress contribution. (After Tritton, 1988.)

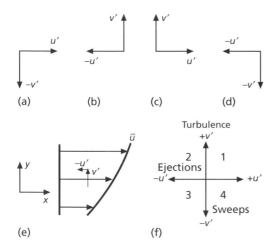

Fig. 5.13 (a)–(d) Geometrical interpretation of Reynolds stress. (After Tritton, 1988.) Patterns (a) and (b) dominate the turbulent boundary layer, giving rise to positive contributions to Reynolds stress. (e) A sketch to show how a Reynolds stress arises from a burst motion in a mean velocity gradient. (f) A quadrant plot to show the four possible types of eddy motion in turbulent flows. Quadrants 1 and 3 contribute little to the turbulent stress.

Similarly for sweep motions, but here the motions are *quadrant 4* events, with faster-than-average downward motion giving the negative product.

Of great sedimentological interest are experimental results (Grass, 1971) on burst/sweep phenomena over boundaries roughened by sediment grains glued one grain thick over the flow channel bed (Fig. 5.14). The use of 2 mm coarse sand and 9 mm pebbles provided transitional and rough boundaries respectively at flow velocities well below the movement threshold for loose grains. The three boundary conditions were analysed using flows of constant Reynolds number. Increasing boundary roughness caused increasing mean bed shear stress, as we would expect. The turbulence intensity data scale directly with u_*, independent of roughness conditions for $h/d > 0.2$, implying that, beyond a certain height, the intensity depends solely on boundary distance and shear stress but is independent of the conditions producing the shear stress. Closer to the bed the data separate so that, with increasing boundary roughness, the longitudinal intensity decreases and the vertical intensity increases. The average Reynolds stress measurements correlate with a linear mean shear stress distribution tending towards zero at the flow free surface.

We can thus envisage the smooth-boundary viscous sublayer fluid and the fluid trapped between the roughness elements as 'passive' reservoirs of low-momentum fluid that is drawn on during ejection phases. Entrainment of this fluid is extremely violent in the rough-boundary case, with vertical upwelling of fluid from between the roughness elements. Very significantly the streaky pattern of the viscous sublayer observed on the smooth boundary is much less conspicuous in the transitional and rough-boundary flows. Faster deceleration of sweep fluid on the rough boundary due to form drag of the grains causes the decrease of longitudinal turbulent intensity and the increase of vertical turbulent intensity noted above. The Reynolds stress contribution ($\overline{u'v'}$) is also increased by increased roughness close to the wall region.

5.7 Shear flow instabilities, flow separation and secondary currents

In many sedimentological contexts fluid flow may occur within or at the base of another fluid. Density currents in the air (like volcanic pyroclastic surges or *nuées ardente* type flows) or under water (like turbidity currents) are examples. In such cases there is shear not only between the moving current and a solid bed, but also between the current and the overlying fluid; we may imagine the situation as like those sketched in Fig. 5.15, with a planar shear layer between two fluids. Now, such a situation becomes unstable if some undulation or irregularity appears along the shear layer, for any acceleration of flow on the part of one fluid will tend to cause a pressure drop and an accentuation of the disturbance. Very soon a striking, more-or-less regular, system of vortices appear (Fig. 5.16) rotating about approximately stationary horizontal axes with respect to the plane of shear. Such vortices (termed Kelvin–Helmholtz instabilities) are highly important mixing mechanisms (and the cause of skin friction) at tributary junctions (Plate 2), on the lee side of dunes and on the fronts and tops of density currents.

The phenomenon of flow separation is very common in natural flows of sedimentological interest, particularly turbulent flows. It is a consequence of the slowing down of fluid as it moves past a boundary. Should the pressure increase in the downflow direction then the near-bed fluid may easily be forced

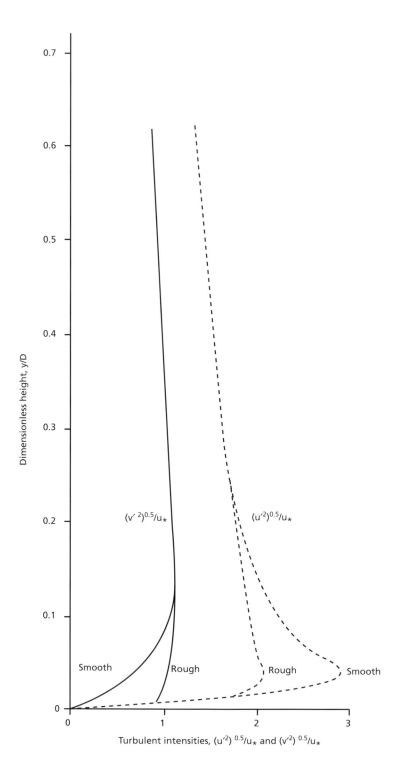

Fig. 5.14 The effect of a rough boundary upon turbulence production intensity and Reynolds stresses in turbulent flows of similar Reynolds number (6700). Note the increase of vertical turbulence intensity ($(v'^2)^{0.5}/u_*$) and the decrease in horizontal turbulence intensity ($(u'^2)^{0.5}/u_*$) from smooth to rough boundaries close to the bed and the independence of intensity on roughness for $h/d > 0.2$. (After Grass, 1971.)

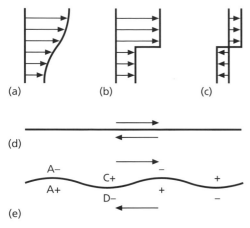

(a)   (b)   (c)

(d)

(e)

Fig. 5.15 Sketches to illustrate the cause of
Kelvin–Helmholtz instability. (After Tritton, 1988.)
(a)–(c) Various velocity distributions that might cause the
instability: (a) is an approximation to a turbulent boundary
layer; (b) and (c) might correspond to fluids in relative
motion such as density-stratified flows. (d) An initially
planar shear layer. (e) A wavy shear layer produced by
pressure differences between (a)/(b) and (c)/(d), etc.
At the convex bumps like (a) and (d) the fluids are
accelerated and the pressure drops; vice versa for
(b) and (c). The system therefore is unstable and deforms
in response to the pressure gradients until some
equilibrium is reached with the viscous resistance
across the boundary.

Fig. 5.16 Beautiful Kelvin–Helmholtz waves produced at
an interface between two immiscible fluids as in the
situation illustrated in Fig. 5.15c (From van Dyke, 1982.)

upstream in a process known as *flow separation*.
Consider flow over a convexity in a bed (Fig. 5.17).
The mean streamlines converge and then diverge.
From the continuity equation (Chapter 4) the flow
will speed up and then slow down downstream.
Bernoulli's equation states that the pressure should
decrease in the accelerated flow section. Pressure and
velocity gradients should therefore exist such that
$dp/dx < 0$ upstream and $du/dx > 0$ downstream. The
pressure gradient so produced has the greatest effect
on the lower-speed fluid near to the bed. This fluid
will be more easily retarded by the adverse pressure
gradient. Under certain circumstances the pressure
gradient will be able to push fluid close to the wall
upstream, as shown in Fig. 5.17. Boundary-layer
separation may take place from point S, named the
separation point, and boundary-layer attachment will
eventually occur downstream. Within the streamline
joining S and the reattachment zone will be a closed,
recirculating separation bubble. Boundary-layer separation also occurs around spheres or cylinders placed
in the flow or when a negative step or small defect
occurs on a bed (Chapter 6).

Both laminar and turbulent boundary layers can
separate, the former usually separating more readily.
However, in sedimentological fields we are most
interested in turbulent separation. We can define two
types of separation zone. Vortex bubbles form when
a downward ramp or step is skewed up to 45° from
the flow direction. Roller bubbles are skewed > 45°
to flow direction. The streamlines of a roller are
closed loops and those in a vortex are helical spirals.
Important effects arise at the upper junction of the

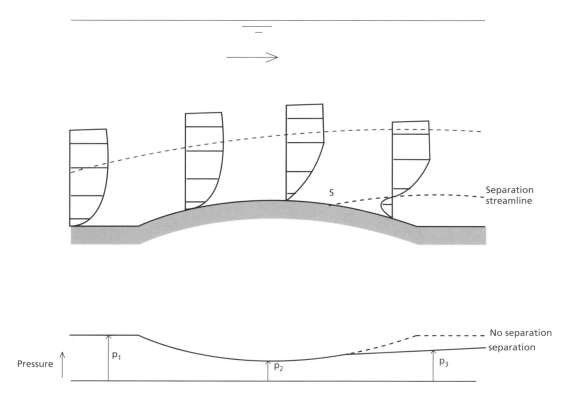

Fig. 5.17 Diagram to illustrate how flow over a convex surface gives rise to an upstream negative and then a downstream positive pressure gradient, the latter causing flow separation. (After Massey, 1979.)

bubbles with the mainstream fluid. Here we have relative motion between two flows, which produces strong vortices of the Kelvin–Helmholtz type along the unstable interface.

Secondary flow is a name given to more-or-less regular patterns of streamlines that diverge from the mean direction of downcurrent or downslope flow. It is common in both straight and curved channels. In straight channels the effect takes the form of streamwise vortex cells showing divergence of flow towards the channel wall and a compensatory flow inwards and across the bottom. Several pairs of these longitudinal vortices may occur. The effect is due to an imbalance of Reynolds' stresses at the wall compared to the main body of the flow, the stress gradients setting up forces that cause the flow pattern observed. Such vortices are responsible for the transport of sand in prominent flow-parallel 'windrows' and for the erosion of flow-parallel 'gutter-marks' (Chapter 10).

Such stable secondary vortical flows are known as Taylor–Görtler vortices. At the bed this is seen as periodic outward and inward transverse flow components. The vortices set up their own transverse pattern of greater and lesser bed shear stresses that cause longitudinal, flow-parallel bedform elements to form (see Chapter 7).

A second type of secondary flow occurs ubiquitously in channel bends; we shall discuss this fully in Chapter 17. There are also several other types of natural secondary flows, due to spanwise changes in roughness for example. These are briefly discussed in a sedimentological context by Pantin *et al.* (1987).

Further reading

The joys of turbulent flows receive a concise introduction from P. A. Allen (1997; Chapter 4). Turbulent structures are illustrated in van Dyke (1982). Clifford *et al.* (1993) and Ashworth *et al.* (1996) are research monographs with many sedimentologically relevant papers on turbulence. Once more, Tritton (1988) will set you on the right path for the mathematical physics of turbulence.

6 Sediment grains in fluids: settling, transport and feedback

Full from the rains, but the flood sediment gone;
Under the brace of the glancing current
Each pebble shines with a life of its own,
Electric, autonomous, world-shaking-divergent.

Hugh MacDiarmid, 'The Point of Honour', *Complete Poems*, Vol. 1, Carcanet

6.1 Introduction

Having established some basic fluid properties and principles of fluid motion, we turn now to the interaction between fluid and sediment grains. In the 1960s R.A. Bagnold gave the name 'loose-boundary hydraulics' to this field of study, distinguishing it from the analysis of pure fluid motion alone. We distinguish two types of sediment bed: granular/cohesionless and cohesive. The first type includes all boundaries made up of solid grains that are kept in contact with adjacent grains purely by gravitational effects. The second type applies most commonly to clay mineral aggregates on mud beds where the tiny clay mineral flakes are mutually attracted by electrostatic forces, which may be large compared to gravitational ones. We shall concentrate in this chapter on the former case.

6.2 Fall of grains through stationary fluids

Sediment particles falling through static or very slowly moving water and air masses are common in Nature, from silt in river-mouth plumes to dead (and dissolving) pelagic shells sinking through the oceans. If we introduce sediment grains of density σ singly into a static liquid of density ρ such that $\sigma > \rho$, then the spheres will initially accelerate through the fluid, the acceleration decreasing until a steady velocity known as the terminal or fall or Stokes velocity (U_S) is reached. A typical graph of experimental data for smooth quartz spheres in water is shown in Fig. 6.1, showing clearly that the fall velocity increases with increasing grain diameter but that the rate of increase gets less.

Accurate prediction of the rate of fall of solids through stationary fluids is only possible for one

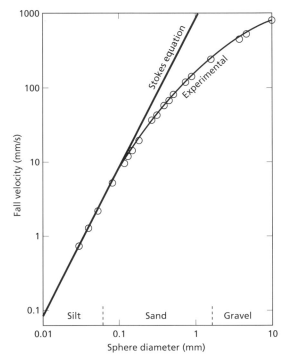

Fig. 6.1 Fall velocity as a function of grain diameter for quartz spheres in water at 20 °C compared to the predictions of Stokes' law. (After Gibbs *et al.*, 1971.)

highly specialized case—that of the *steady* descent of *single, smooth, insoluble spheres* through a *still Newtonian fluid* in *infinitely wide and deep* containers when the *grain Reynolds number is low* (< 0.5) and hence viscous flow separation does not occur. The prediction was first made by G. Stokes, the 19th-century English physicist. The reader should have noted the eight or more restrictions to the Stokes analysis (*italicized* above), perhaps concluding that

Further information on
Stokes' law

There are two ways of approaching this problem. An engineer's or experimentalist's approach is to group together the variables that might control the velocity of fall and to do dimensional analysis. Thus we might surmise that the fall velocity will vary directly with density contrast, $\sigma - \rho$, size as diameter or radius, d, and gravity, g. We might expect it to vary inversely with fluid viscosity, μ. Thus we could write a proportionality relationship of the general form:

$$U_S \propto \frac{(\sigma - \rho)dg}{\mu} \tag{B6.1}$$

This leaves the determination of the proportionality constant and any nonlinear behaviour to experimentation. The approach of the mathematical physicist (originally G. Stokes) is to solve the problem theoretically by recourse to the fundamental equations of flow and potential flow theory, both topics we have lightly touched upon in Chapters 4 and 5. Stokes considered viscous fluid resistance forces only and arrived at a theoretical relation that expresses fall velocity as a function of grain and fluid properties. During steady fall, all acceleration terms in the equations of motion vanish, leaving a balance between the solid grain's immersed weight force F_w and the viscous drag force F_v acting over the solid grain surface. (Derivation of F_v from first principles is a lengthy business and beyond the scope of this book. Faber (1995) gives a good account, although the reader will need some higher mathematics to understand it.) We now have:

$$F_w = F_v \tag{B6.2}$$

or

$$\tfrac{4}{3}\pi r^3(\sigma - \rho)g = -6\pi\mu r U_S \tag{B6.3}$$

where r is the grain radius, μ is the molecular viscosity and U_S is the fall velocity. This may be rearranged to find the fall velocity, frequently called the Stokes velocity in honour of its discoverer:

$$-U_S = \frac{2}{9}\left(\frac{r^2(\sigma - \rho)g}{\mu}\right) \tag{B6.4}$$

Note that U_S changes sign for buoyant systems where $\sigma < \rho$. As noted previously Stokes' law only accurately predicts the fall velocity of particles whose particle Reynolds number (where the velocity term is the full veocity) is < 0.5. This corresponds to silt-sized and finer, quartz-density particles in water. At higher grain Reynolds numbers the viscous resisting force $-6\pi\mu r U_S$ becomes an underestimate, as marked accelerations of the bounding fluid and flow separation effects (Fig. B6.1) begin to occur. No satisfactory theoretical solution is now possible, so recourse has to be made to experimental data on rates of fall and on drag coefficients. In the Stokes range we may equate the two forms of expression for the surface drag force. Thus:

continued on p. 125

Further information on Stokes'
law [*continued*]

Fig. B6.1 Visualization of flow around a sphere to show the onset of turbulent flow separation. (From van Dyke, 1982.)

$$6\pi\mu r U_S = C_D \pi r^2 \rho \frac{U_S^2}{2} \tag{B6.5}$$

or

$$C_D = \frac{24}{Re} \tag{B6.6}$$

for the drag coefficient. Outside the Stokes range C_D must be gained from experiment.

the formula is practically useless. Thus natural silicate grains or calcareous biological debris are not spherical, usually fall in a group and the latter are soluble. Nevertheless, we may make use of Stokes' formula as a good approximation in some natural situations.

In many natural systems, other problems arise because of the presence of rough, nonspherical particles and multigrain settling. The problem of non-sphericity must be solved by recourse to specific experiments, particularly when fragments of biological grains at high grain Reynolds numbers (Re_g) are involved. In multigrain settling the fluid streamlines relative to the individual grains interact. Increased drag then results in a decrease of the grain fall velocity relative to its velocity in a grainless fluid. It has been proposed (Richardson & Zaki, 1958) that the fall velocity, U_S', of a spherical particle in a dispersion of other falling grains varies as

$$U_S' = U_S(1 - C)^n \tag{6.1}$$

where U_S is the fall velocity of a single grain in an otherwise grainless fluid, C is the volume concentration of grains in the falling dispersion, and n is an exponent varying between 2.32 (for $Re_g > 1$) and 4.65 (for $Re_g < 1$). This relation tells us that the fall velocity of a grain in a dispersion will be smaller than that in an otherwise grainless fluid and strongly dependent upon concentration. For fine sediment, when $n = 4.65$, and at a high value of C around 0.5, U_g' may be only a few per cent of U_g. This consideration is of great importance in understanding the settling behaviour of dense, sediment-laden flows such as turbidity currents. Further interesting complications arise:

• when a grain is soluble such that its size and mass decrease (but not necessarily linearly) during fall;
• when a grain falls through a density-stratified fluid;
• when a grain falls through or into a Bingham fluid whose yield strength may exceed the applied gravitational force due to the grain itself.

Each of these situations is common in Nature. Further examples will doubtless spring to the reader's mind.

6.3 Natural flows carrying particulate material are complex

The sedimentologist studies natural flows and the majority of these are complex in that they contain various types of particulate matter that influence fluid behaviour under conditions of shear. Three end-members may be defined, although natural 'dirty' flows may feature several kinds of behaviour:

1 In fresh water or seawater transporting cohesionless, non-electrically-charged grains of silicate minerals, additional resisting stresses may arise through grain-to-grain interactions, leading to an increase in the apparent viscosity of the two-phase system (Bagnold, 1956, 1966b).

2 In freshwater flows transporting cohesive particles, Bingham-like behaviour occurs, with significant increases in apparent bulk viscosity directly related to the amount of transported clays and rate of shear (Wan, 1982). Such behaviour is to be expected if the cohesive particles behave as rigid aggregated particles analogous to colloids. The fluid must flow around the rigid aggregate rather than through it and hence the viscous dissipation of the flow must increase (Witten, 1990).

3 Seawater flows transporting dilute (0.15–10 g/L), flocculated, organic-rich aggregates of clay minerals have been found to maintain a Newtonian flow structure and show no noticeable gradient of sediment concentration towards the wall (Gust, 1976). The thickness of the wall layer may be enhanced by a factor of 2–5 and the friction velocity decreased by up to 40%. This drag-reducing behaviour is the opposite of that expected if the clay particles had caused a small increase in fluid viscosity and has been postulated to be a naturally occurring analogue of plain water flows to which have been added small amounts of polymer molecules (Gust, 1976; Best & Leeder, 1993).

Any reduction in fluid drag exerted adjacent to the bed by smooth boundary flows must come about by decreases in some or all of the following:

• apparent molecular viscosity;
• velocity gradient in the viscous sublayer;
• Reynolds stress production.

The first effect is the opposite to that expected if small amounts of clay are added to a flow since a small increase in molecular viscosity will result. The other causes of drag reduction have been widely used to manipulate turbulent boundary layers. Studies with drag-reducing polymers (the Toms effect; e.g. Tiederman et al., 1985) have revealed that drag reduction is achieved through modification of the near-wall turbulence structure within the buffer region of the flow, and that the viscous sublayer plays a relatively passive role. This indicates that drag reduction is inherently linked to changes in the mechanisms of turbulence production within the buffer region; turbulent momentum exchange is evidently reduced in subsequent interactions with the outer regions of the boundary layer.

6.4 Fluids as transporting machines

The transport of sediment grains by shearing fluid must be due to exchange of momentum between grain and fluid, i.e. forces are set up during the transport

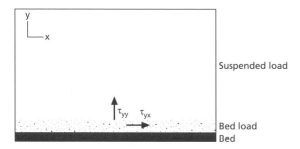

Fig. 6.2 Simple definition sketch to show the two stresses responsible for bedload and suspended-load transport.

process. Two components of the stress tensor τ_{ij} are chiefly involved: τ_{yx} drives the transport of grains close to the bed via the production of lift and drag forces; τ_{yy} supports the load in the remainder of the flow. Figure 6.2 illustrates this division of labour.

Working from fluid dynamic principles as discussed in earlier chapters, we might expect the following:

1 In order to move a layer of stationary grains as bed-load, it must be sheared over the layer below. This process involves lifting the immersed mass of the top-most layer over the underlying grains as a dilation, and hence work must be done to achieve this result.

2 The energy for the work of bedload transport must come from the kinetic energy of the shearing fluid boundary layer; whether the fluid is in a state of laminar or turbulent flow is immaterial.

3 Close to the bed it might be expected that the fluid momentum transferred to the moving grains will in turn be transferred to other stationary or moving grains so that a dispersion of colliding grains will eventually evolve. The efficacy of grain collisions will depend upon the immersed mass of the grains, the dynamic friction coefficient and the viscosity of the moving fluid. We will thus expect major differences between wind and water transport.

4 If grains are to be transported in the body of the fluid, then some fluid mechanical mechanism must act to effect the transfer of grains from the bed layers. This mechanism must be sought in the processes of turbulent shear, chiefly in the burst and sweep motions outlined in Chapter 5.

The fact that fluids may do useful work is obvious from their role in powering waterwheels, windmills and turbines. In each case the kinetic energy becomes the mechanical energy of the machine in question.

Energy losses occur, with each machine operating at a certain efficiency as a result. For each case:

work rate = available power × efficiency

Applying these basic principles to a river or delta channel—perhaps the simplest natural systems to begin with—the concept of useful work is now replaced by natural work. The river will try to transport the sediment grains supplied to it by hillslope processes, tributaries and bank erosion. How much sediment can be carried? This will depend upon the power available to the river, the local power supplied to the channel boundaries and the efficiency of the energy transfer between fluid and grain. As we saw in Chapter 5, flow power is made available as the fluid potential energy is converted to kinetic energy down a gravity slope. This power is available to erode and transport sediment grains. However, as with any machine, only some portion of this power is utilized in the mechanical process of doing work, i.e. the river may be said to operate at some efficiency, e.

Useful as the concept of flow power is for channel-ized water flows, there are problems in other systems:

1 The wind has no readily definable upper flow limit.

2 The availability of sediment is not uniform, and it depends upon delivery.

3 Nonequilibrium effects may be very important. Clays or very fine silt grains have settling velocities so low that they may be carried far downstream by a flow before they can settle to the bed. Thus the flow may be said to be oversaturated with sediment.

6.5 Initiation of particle motion

As fluid shear stress over a levelled plane bed is slowly increased, there comes a critical point when grains begin to be moved downstream with the flow. Sediment transport has begun. A particular fluid shear velocity above the threshold for motion may be expressed as a ratio with respect to the critical threshold velocity (u_{*C}) for the grains in question. This is the transport stage (Francis, 1973), defined as the ratio u_*/u_{*C}.

A great deal of attention has been paid to the determination of the critical threshold for grain movement since it is an important practical parameter in civil engineering schemes (canals, irrigation channels, model experiments). A knowledge of the threshold value for different grain types and sizes is also of

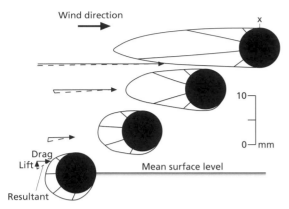

Fig. 6.3 The pattern of approximate pressure differences measured in experiments between position X on top of a 7.5 mm sphere and other positions on the sphere at various heights in a windstream. The lengths of the lines in the shaded areas outside the spheres denote the relative differences in air pressures. Both lift and drag forces act on the sphere, but lift decreases rapidly with height whereas drag increases because of the direct pressure of the wind. The wind velocity at 20 mm above the surface in the experiments was 7.7 m/s; shear velocity was 0.98 m/s. (After Chepil, 1961.)

Grains of silt, sand or gravel comprising natural sediment beds are acted upon by a pervasive surface shearing stress as any fluid, be it laminar or turbulent, moves over them. If τ_0 is the mean bed shear stress, then the mean drag per grain is given by τ_0/n, where n is the number of particles over unit bed area. A lift force due to the Bernoulli effect also exists. Fluid streamlines over a projecting grain will converge, the velocity will increase and therefore, to maintain the energy equilibrium, the pressure must decrease above the grain. Chepil's (1961) (Fig. 6.3) and later results leave no doubt that the lift force is comparable to the drag force when the grain is on the bed. The lift force rapidly dies away and the drag force rapidly increases as the grain rises from the bed.

Both fluid forces considered above will try to move bed grains; they are resisted by the grain's normal weight force. A useful way of nondimensionalizing these threshold forces is to express the ratio of applied shear to normal weight forces, grain motion occurring when the fluid force exerts a critical moment about the pivot point of the stationary grain with its neighbour (Fig. 6.4).

It is difficult to determine theoretically the critical applied shear stress for grain motion, despite the initial attractions of using this 'moments of force' approach. This is because there are a large number of variables involved, one of the most unpredictable

special interest to the sedimentologist interested in determining conditions when erosion or deposition will occur and in establishing estimates for the magnitude of ancient currents.

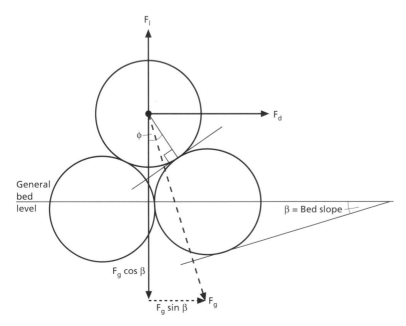

Fig. 6.4 Force moment balance diagram for the entrainment of a single sediment grain pivoting at angle ϕ about an adjacent like-sized grain on a bed sloping at angle β. F_d, F_l and F_g are the drag, lift and gravity forces. At equilibrium, $F_g \sin \beta + F_d = (F_g \cos \beta - F_l) \tan \phi$ See Bridge and Bennet (1992) for exhaustive discussion.

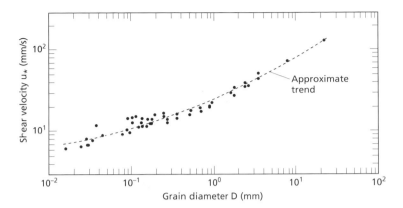

Fig. 6.5 The variation of threshold flow velocity 1.0 m above the bed necessary for the initiation of movement of quartz-density grains in water at 20 °C. (Modified after Miller *et al.*, 1977, by approximate trend.)

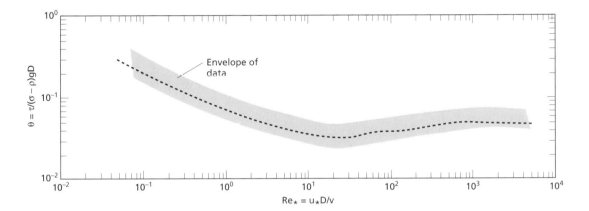

Fig. 6.6 The variation of threshold nondimensional shear stress vs. grain Reynolds number for water flow at 20 °C over a wide range of quartz-density grains. (Modified after Miller *et al.*, 1977, by envelope of data.)

being the degree of exposure of individual grains, no matter how carefully a flat bed is prepared. It is also difficult to estimate the lift force contribution. The critical conditions for the initiation of particle motion must therefore be determined experimentally. The simplest plot involves just a measure of flow velocity vs. grain diameter (Fig. 6.5), but, in order to have the greatest generality, experimental results must be applicable to a wide range of fluids and particles. As sedimentologists we are most interested in natural mineral grains in air and water, but these systems must be treated as special cases of a more general application. Working from first principles we might expect the critical conditions for particle motion, C_c, to be dependent upon gravity (g), grain size (diameter

or radius), immersed weight ($\sigma - \rho$), fluid kinematic viscosity (ν) and bed shear stress (τ_0). Thus:

$$C_c = f(d, g, (\sigma - \rho), n, \tau_0) \qquad (6.2)$$

Now, great generality will result if we arrange these quantities into dimensionless groups for the purpose of plotting experimental results:

$$\theta_c = \frac{\tau_0}{gd(\sigma - \rho)} = f\left(\frac{u_* d}{\nu}\right) = f\left(\frac{D}{\delta}\right) \qquad (6.3)$$

where θ_c is the critical dimensionless bed shear stress. As we surmised previously, this is the useful ratio of applied shear stress to resisting grain stress. The third expression is a grain Reynolds number, Re_g. A plot of θ vs. Re_g (for liquids, known as a Shields diagram after the German engineer A. Shields) is shown in Fig. 6.6.

Although considerable scatter is evident in Fig. 6.6, because of the many different sets of experimental conditions and the difficulty in deciding exactly when the threshold is reached, it appears that θ is nearly

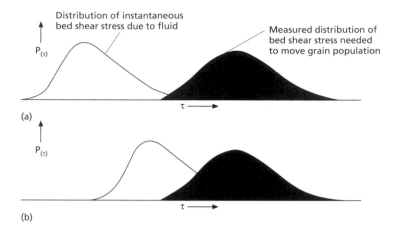

Distribution of instantaneous bed shear stress due to fluid

Measured distribution of bed shear stress needed to move grain population

(a)

(b)

Fig. 6.7 Graphoids to show that the threshold of grain movement must be defined by the statistical overlap between the distribution of instantaneous applied shear stresses due to the turbulent fluid and the actual distribution of stresses needed to move the particular grain population. (a) Small overlap, little motion. (b) Larger overlap, general threshold exceeded. (After Grass, 1970.)

constant, at an average of about 0.05, for a wide range of grain diameters up to a grain Reynolds number of about 1.0. At low grain Reynolds numbers θ increases steadily to around 0.3. This increase is thought to be due to the presence of a smooth boundary to the flow (Chapter 5), with the particles lying entirely within the viscous sublayer. Here the instantaneous velocity gradients are less than in the lowest part of the turbulent boundary layer.

Important light has been thrown on the 'threshold problem' in water by regarding the onset of grain movement as resulting from the interaction between two statistically distributed variables (Grass, 1970). The first variable may be termed the initial movement characteristic of a given bed material in a fluid of given viscosity and density. Thus every grain on the bed is assumed to be susceptible to a local instantaneous stress, and because of the random shape, weight and placement of individual grains this τ_0, termed τ_c, has a probability distribution. The second variable is the local instantaneous bed shear stress caused by burst/sweep events (Chapter 5). These stresses also have a probability distribution, dependent on τ, fluid density, viscosity and flow boundary conditions. At the onset of grain motion the most susceptible particles (those with the lowest characteristic critical shear stress) are moved by the highest shear stresses in the shear stress distribution applied to the bed by the flow. Experimental results enable histograms to be constructed for each distribution (Fig. 6.7). These show that the critical shear stress necessary to move grains occurs when the two τ distributions overlap by a certain constant amount. Much of the scatter on the Shields plot evidently reflects different observers deciding on different degrees of overlap.

Perhaps the most intractable aspect of the 'initiation of motion problem' is the fact that many natural sediment beds are of radically mixed sizes, perhaps bimodal or even polymodal in their size distributions. A lot of experimental and theoretical research has gone into trying to come up with general expressions to cope with this common situation (Parker *et al.*, 1982; Wilcock & Southard, 1989; Paola *et al.*, 1992; Bridge & Bennett, 1992). It was initially pointed out that, for a pebbly sand, the motion of the pebbles occurred at lower critical applied fluid stress than for pebbles alone, because of the effects of greater exposure of the pebble to the shearing boundary layer and the tendency for the sand around the pebble to be scoured, thus elevating the pebble into the boundary layer and transporting sand at the same time. This has been named the condition of 'equal mobility' at threshold. However, the situation rarely reaches equilibrium because close to threshold the bed quickly 'armours' once a certain amount of the sand has been so scoured away, that is, a surface layer of interlocking and immovable pebbles protects the underlying sand–pebble mixture. It is only well above the threshold that equal mobility conditions seem to apply. The process of upstream armouring and a consideration of the great increase in turbulence intensity, and hence potential for suspension of sand, over mixed-size rough beds also serves to explain the almost universal trend found in rivers for downstream fining (see Paola *et al.*, 1992).

6.6 Initiation of motion by air flow

In air flows Bagnold (1954b) defined two types of threshold. At a critical value of air speed insufficient to move bed grains by fluid shear alone, grain motion could be started and propagated downwind by simply letting sand grains fall on to the bed. Other grains were bounced up into the airstream, which, upon falling, caused further movement as they impacted on to the bed, and so on. Grain motion ceased as soon as the introduction of artificial grains stopped. The critical wind speed necessary for this process was termed the *impact threshold*. Further increase of air speed enabled grains to be moved by the direct action of the wind at the fluid or *normal threshold*. The ability of natural sand grains in air to disturb and eject other grains after impact contrasts with the situation in water. This is because of the viscosity contrast between air and water, which controls resistance to motion, and the great effective density contrast between quartz and air (2000 : 1) compared to quartz and water (1.65 : 1). As we shall see later (Chapters 7 & 8), the types of ripples developed in the two fluids depend intimately upon these factors. For most sands in air flows the critical shear velocity is a function of the square root of the particle size.

6.7 Paths of grain motion

Once in motion, other contributions to fluid surface forces exist. Shear lift occurs due to relative grain motion through a velocity gradient. Spin lift occurs when a grain undergoing relative motion in a fluid rotates—the Magnus effect is responsible. Once the threshold for motion is exceeded, grains move downstream in three basic ways (Fig. 6.8).

1 *Rolling.* Rolling motion is simply defined as continuous grain contact with the bed and it includes the rarely observed 'sliding' motion. In air, rolling grains form a 'creep carpet' kept in motion by the kinetic

Fig. 6.8 Cartoon to show the various ways that grains are transported in water and air flows.

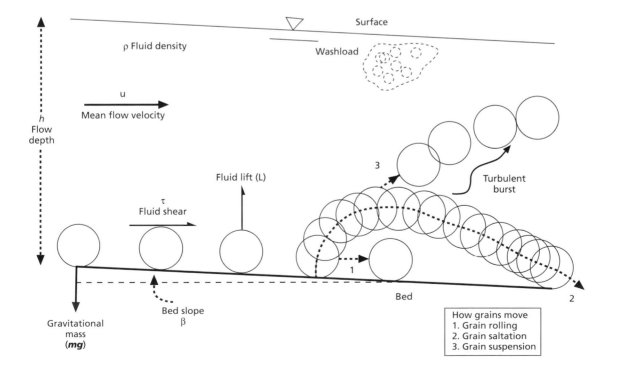

energy exchanged by the impacts of incoming saltating grains.

2 *Saltation.* In saltating motion (Latin *saltare*, to dance) grains rise steeply (> 45°) from the bed, ascending to a height of a few grain diameters, then undergo a smooth motion like a ballistic trajectory leading into a shallow-angled (> 10°) descent back to the bed. The trajectory is not usually the shape of a symmetrical parabola as shown, say, by the trajectory of an artillery projectile, a thrown stone or an arrow. Saltation may occur in both laminar or turbulent flows. In air, because of the very high density contrast between air and quartz and the low viscous resistance of the air, it is the collision process that causes initial upward motion of other grains. In water, collisions are dampened by viscous retardation after the grain's elastic rebound has occurred—the grain momentarily halts or undergoes a rolling motion (i.e. grain linear momentum is not conserved) before fluid forces make it rise once more. Grain spin during flight causes Magnus effects, which probably aid particle ascents and steepen descents.

3 *Suspension.* Suspended (literally 'held up') motion involves grains moving in generally longer and more irregular trajectories higher up from the bed than in saltation. A saltating grain may experience an upward acceleration during the descending part of a trajectory as turbulence bursting just starts to affect saltation—a state of incipient suspension.

It is often difficult to follow the path of an individual moving sand grain, particularly in a crowd of other grains, and it is only from the application of skilful photographic techniques that we have a full idea of the different types of grain paths. Multi-exposure photographic techniques (Francis, 1973; Abbott & Francis, 1978) show that:

- the varying proportions of time spent in the above three modes by grains in water are a direct function of transport stage (defined in Section 6.5);
- the statistics of the increase of length and height of grain trajectories are direct functions of transport stage;
- the mean forward speed of grains is a direct function of the mean flow velocity.

The onset of fully developed suspension may be roughly estimated by assuming that the root-mean-square values of the upward vertical velocity fluctuations close to the bed begin to exceed the fall velocity of the saltating grains, i.e.

$$U_S \leq \sqrt{\overline{v'^2}} \qquad (6.4)$$

Since it is known that the term involving v' reaches a value of about $1.2u_*$ at maximum, we can write an approximate criterion for full suspension as:

$$U_S/u_* \leq 0.8 \qquad (6.5)$$

This criterion successfully explains the onset of full suspension effects observed in multi-image photographic experiments. It might be thought from this criterion that fine bed sediments ($d < 0.1$ mm) should become suspended immediately the threshold conditions for motion are reached. The fact that this does not occur (current ripple bedforms evolve on the bed) may be attributed to the burial of the tiny grains deep within the viscous sublayer, protecting them from the action of suspensive burst motions.

The characteristic grain paths defined above apply only to grain transport when grains have no effects upon each other during movement, i.e. when simple saltant trajectories are not interrupted by grain–grain collisions or deflections. Such interference effects are more likely in water than in air because of the thinness of the saltation zone in the former case. As the transport stage of water flows is increased, more and more grains are entrained and hence grain concentration increases. Grain–grain collisions become certain. It can be shown by simple collision dynamics based upon kinetic theory that simple saltations will cease to exist over a transport stage of about 2 when the grains close to the bed begin to move as a concentrated granular dispersion dominated by grain–grain impacts or deflections.

We have now reached a suitable point to give rational definitions of the types of transported sediment loads (Fig. 6.9).

1 *Bedload* (traction load) includes rolling, saltating and collision-interrupted 'saltating' grains. The grains comprising bedload transfer momentum to the stationary bed surface by solid–solid contacts and the rate of change in this momentum over unit area in unit time must equal the immersed weight of the bedload grains. Our definition of bedload is thus both positional and dynamic (Bagnold, 1973).

2 *Suspended load* includes all grains kept aloft by fluid turbulence so that the weight force of the suspended grains is balanced by an upward momentum transfer from the fluid eddies. The process is more efficient in water flows than air flows because of the

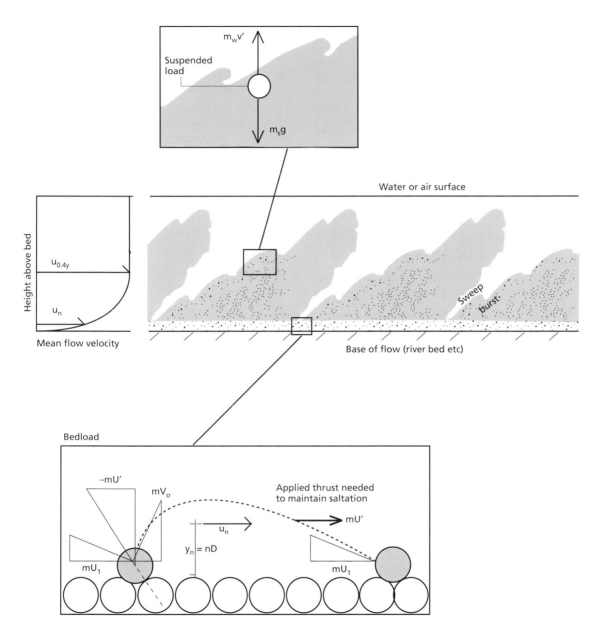

Fig. 6.9 Diagrams to illustrate the division of sediment load into bedload and suspended load. The shaded fluid in the upper two sketches represents bursting boundary-layer fluid moving upwards in the flow carrying suspended sediment along with it. The lower sketch shows the applied thrust needed to maintain saltation. m_w, m_s, water and grain mass; v', vertical turbulent velocity; u, flow velocity; u_n, effective velocity at centre of thrust for bedload; y_n, height of u_n above bed; U_1, incoming grain velocity; V_0, outgoing grain velocity; U', grain velocity lost on impact. (After Bagnold, 1973.)

high density contrast between grain and fluid and low fluid viscosity.

3 *Washload* is a broad term used to describe the more-or-less permanently suspended clay-grade 'fines' present in water flows. *Dustload* is a more suitable equivalent term for air flows. As noted previously in the context of oversaturated flows, fine particles would fall back to the bed fairly slowly if the flow were to cease.

The transport of bedload sediment may have marked effects upon the distribution of fluid velocity with height from the bed surface. A clear example is provided by the wind case, where velocity measurements are possible within and just above the bedload zone using a very fine pitot tube (Bagnold, 1954b).

As the wind strength increases above the threshold for movement, there is a clear retardation of air velocity in the centre of gravity of the bedload zone. The retardation reaches about 20% of the pure air flow velocity expected by the Karman–Prandtl equation (Chapter 5) for turbulent flow in the absence of moving solids. The departure of the measured velocity profiles from the straight line is explicable by the expected transfer of momentum from fluid to solid over the upper, accelerated, part of the saltant trajectories. A similar effect should be present in water flows (Bagnold, 1973), but it is difficult to measure because of the thinness of the bedload zone in this case.

Further information on subaqueous bedload transport

Whatever the form of bedload motion, the immersed weight of grains present over unit bed area in a steady uniform flow must be in equilibrium with an equal and opposite stress due to the momentum exchanged by periodic contacts of the grains with the bed. In order to overcome frictional resistance at successive contacts, it is necessary to apply an external impulse to the grains from the fluid. The magnitude of all forces involved will obviously depend upon the net mean applied fluid shear stress, τ_{yx}. The bedload immersed weight, w_b, may be written as:

$$w_b = m'_b g \tag{B6.7}$$

where m'_b is the bedload immersed mass per unit area. This is in equilibrium with a *solid* transmitted normal reaction, T_{yy}. From basic considerations of friction:

$$T_{yy} \tan \alpha = T_{yx} \tag{B6.8}$$

where $\tan \alpha$ is a dynamic friction coefficient and T_{yx} is the transverse, bed and flow-parallel component of T_{yy}. Now T_{yx} must be balanced by an equal and opposite fluid force, given by the total applied fluid shear stress minus the critical stress needed for the initiation of bedload motion. In symbols:

$$(\tau_{yx} - \tau_{yx,\text{crit}}) = T_{yy} \tan \alpha = T_{yx} = m'_b g \tan \alpha \tag{B6.9}$$

Returning to Bagnold's concept of flowing fluid as a transporting machine, where

$$\text{work rate} = \text{available power} \times \text{efficiency} \tag{B6.10}$$

the work rate is given by:

$$\text{work rate} = (m'_b g \overline{U_b}) \tan \alpha \tag{B6.11}$$

continued on p. 135

Further information on subaqueous bedload transport [*continued*]

where the bracketed term gives the transport rate, i_b, and $\overline{U_b}$ is the bedload transport velocity. The available power is given by:

$$\text{available power} = \frac{k(\tau_{yx} - \tau_{yx,\text{crit}})\overline{u_b}}{\tan \alpha} \qquad (B6.12)$$

where k is a coefficient (< 1) determining the part of τ_{yx} that is transferred to the bed indirectly via the saltating solids and $\overline{u_b}$ is the fluid velocity in the bedload zone. The efficiency is given by:

$$\text{efficiency} = \frac{\overline{U_b}}{\overline{u_b}} \qquad (B6.13)$$

and is thus a function of the 'slip' or relative velocity of solids and fluids in the centre of gravity of the bedload zone. So, finally, we have the complete bedload transport equation:

$$i_b = \frac{k(\tau_{yx} - \tau_{yx,\text{crit}})\overline{U_b}}{\tan \alpha} \qquad (B6.14)$$

This equation can only be solved with an empirical knowledge of k for different transport stages and bed states. Thus far k has proven difficult to estimate in any rigorous fashion. Experimental values for $\tan \alpha$ and for $\overline{U_b}$ are better established, but there is still considerable doubt as to the general applicability of the data. Equation (B6.14) may only be rigorously tested for cases where there is no appreciable differential bed relief and thus no flow separation contribution to the bed shear and where intense sediment transport by bedload alone causes $k = 1$ in a concentrated dispersion. The restrictive nature of eqn (B6.14) to anything other than strictly controlled laboratory conditions has led to the almost universal use of empiricism in sediment transport studies. It appears that after more than 60 years of serious scientific study the majority of natural transporting systems must be measured and sampled in order to obtain any reliable indication of transport rate.

6.8 Solid transmitted stresses

From the discussion above it is implicit that, once grains are in motion as bedload, impacting grains will transfer momentum from flow to the bed and give rise to additional solid transmitted stresses. For example, consider a saltating grain of mass m impacting on to a bed with velocity U and making an angle ζ with the bed surface. If elastic rebound occurs, and neglecting frictional effects, then momentum of magnitude $2mU \sin \zeta$ is transferred normal to the bed surface and momentum $2mU \cos \zeta$ is transferred tangential to the bed surface. If no rebound occurs and a period of rolling ensues after impact, then this momentum transfer is halved.

It is difficult to measure directly the magnitude of the solid stresses T and P due to the rate of change of tangential and normal momentum, respectively, in flow channel experiments (but see Bagnold, 1955). In an ingenious experiment using a rotating coaxial drum, Bagnold (1954a) was able to measure the magnitude of T_{yx} and the normal transmitted stress T_{yy}. The results of the experiment reveal that there is a great increase in the shear resistance of the solid–fluid mixture when compared to the plain fluid alone. Two regions of behaviour were definable in terms of a dimensionless number now known as the *Bagnold number*. A viscous region of behaviour at low strain rates and/or low grain concentrations showed the ratio T_{yx}/T_{yy} to be constant at around 0.75. In this

For the case of bedload transport in wind we cannot use the power analogue as in the water flow case since the height of a wind is indeterminate. Bagnold's (1954b) approach in this case is to equate the rate of applied shear by the moving air to the rate of loss of momentum by the bedload grains. Consider a grain of mass m moving from rest and striking the bed at velocity U after travelling a distance L. If all the velocity U is lost at impact, then the grain may be considered to have gained momentum of magnitude mU/L per unit length of travel from the air. Now if a mass, i_b, of equal-sized grains of sand in saltation moves along a lane of unit width and passes a fixed point in unit time, then the rate of loss of momentum will be $i_b U/L$. The rate of loss of momentum is a force so that the expression $i_b U/L$ is a measure of the resistance exerted on the air over unit surface area due to grain saltation. Calling the resisting or shear force τ, or, from Chapter 5, ρu_*^2, we can write:

$$\tau = \rho u_*^2 = i_b U/L \tag{B6.15}$$

Now it has been found that U/L is equal to g/V where V is the initial vertical velocity of the saltating grain. Hence:

$$\rho u_*^2 = \frac{i_b g}{V} \tag{B6.16}$$

or

$$i_b = \frac{\rho u_*^2}{g} V \tag{B6.17}$$

V should be proportional to u_* so that $V = k u_*$ where $k = 0.8$ from experiments. Substitution yields:

$$i_b = \frac{\rho u_*^3}{g} 0.8 \tag{B6.18}$$

Allowing for the 25% of bedload moved as creep by impact from saltating grains, we arrive at the final expression:

$$i_b = 1.1 \frac{\rho u_*^3}{g} \tag{B6.19}$$

which gives reasonable agreement with experimental data. Note the cubic form of eqn (B6.19). Sediment transport rate is thus strongly affected by increasing flow velocity. Bagnold calculates, for example, that a strong wind blowing at 16 m/s (35 mph) will move as much sand in 24 hours as would be moved in three weeks by a wind blowing steadily at 8 m/s (17.5 mph). We should note, however, that the net long-term transport rate is dominated by moderate winds, since these make up for their low magnitude by having high frequencies.

Further information on bedload sediment transport equation for wind

region grain–grain effects were caused by near approaches causing a repulsion before solid collision could occur. An inertial region of behaviour at high strain rates and/or high grain concentrations showed the ratio T_{yx}/T_{yy} to be constant at around 0.32. Grain–grain collisions dominated in this region. Applying these results to natural flows, Bagnold postulated that:

1 In air, owing to the 'chain reaction' process described above at threshold, so much fluid momentum is transferred to the saltation that virtually the whole of the applied shear stress is resisted by it.

2 In water, where the density ratio of solid to fluid is low, the contribution of T_{yx} gradually increased with transport stage. When one whole, formerly stationary grain layer was in motion as bedload at high transport stage, Bagnold argued that $\tau = T_{yx}$.

3 The normal stress T_{yy}, also termed the dispersive stress, should be in equilibrium with the normal stress due to the bedload weight over unit bed area.

6.9 A dynamic sediment suspension theory

To suspend something is literally to hold it up. A mass of sediment grains must be held above a bed by the eddies of fluid turbulence in a state of suspension. Applied to individual grains, the concept of suspension is statistical, because of the continuous exchange of grains between bedload and the overlying turbulent flow. However, a steady state exists with respect to a suspended mass in a steady, uniform flow. Thus over a sufficiently long period of time the measured mass will itself be constant whereas the constituent grains that define the mass may continuously be exchanged between bed, bedload and suspended load. A state of dynamic equilibrium thus exists. Two questions follow, first formulated by Bagnold (1966b):

• What mechanism acts to keep up the suspended grains?
• What dynamic quantity controls the magnitude of the suspended mass?

As Bagnold predicted, experimental studies and theoretical considerations (Sumer & Oguz, 1978; Sumer & Deigaard, 1979; Leeder, 1983; Wei & Willmarth, 1991) leave no doubt that suspension is achieved by the advection of masses of fluid away from the general region of the bed into the turbulent boundary layer. These burst motions essentially siphon off masses of

buffer-layer fluid and their contained bedload grains into the turbulent flow. Decay of upward-directed fluid momentum in these bursts then causes grain fallout towards the bed, whereupon the process may begin again in response to further burst motions. Grain return to the bed may be aided by 'sweep' motions, whereby fluid from the buffer layer is carried towards the bed.

This dynamic approach to suspension contrasts with the approach of classical suspension theory as a diffusional process as summarized by Coleman (1981) and McTigue (1981). As a scientific theory, the diffusional approach is based upon the analogy of mass transfer to momentum transfer. Mass transfer may be described by simple diffusional considerations of the form required for equilibrium across any horizontal plane in a flow. Thus:

$$\frac{dC}{dy}\varepsilon_s = -CU_s \tag{6.6}$$

where C is the concentration of suspended grains in mass per unit volume, y is height above the bed, ε_s is the sediment transfer (diffusion) coefficient and U_s is the fall velocity. Manipulation and integration of this equation yields:

$$C = C_a \exp\left(-\frac{U_s(y-a)}{\varepsilon_s}\right) \tag{6.7}$$

where C_a is the concentration of grains at a reference height, a, above the bed, usually taken as the centre of gravity of the bedload zone. Equating ε_s with the turbulent fluid momentum transfer coefficient then yields a solvable equation in which C may be calculated for all heights above the bed in the logarithmic zone of the flow.

The diffusional approach sketched out above successfully predicts the concentration profile observed in turbulent suspensions. The theory fails to explain the fact that a statistically steady mass of suspended grains has an excess immersed weight that must be balanced by an equal and opposite upthrust on the part of the turbulent fluid. McTigue (1981) has made progress in relating classical diffusional theory to this dynamic prerequisite, but his analysis does not require asymmetry of vertical turbulent velocities to occur and hence his theory cannot explain steady suspension and cannot estimate the magnitude of the suspended mass in the absence of data on the

Bagnold's dynamic theory of suspension (Fig. B6.2) starts from the premise that the maintenance of a statistically constant suspended mass requires the production of a residual upward normal stress, τ_{yy}, generated by shear turbulence. In symmetrical turbulence (the conventional view) the RMS mean vertical velocity fluctuation is assumed symmetrical about a mean $\bar{v}$ of zero (i.e. a normal distribution of v values is present). Considering unit volume of fluid, the upward and downward

Further information on Bagnold's dynamic theory of suspension

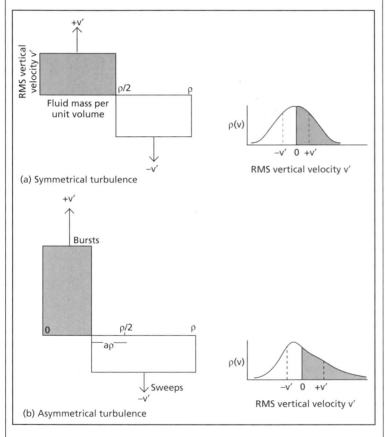

Fig. B6.2 Diagrams to illustrate Bagnold's hypothesis for the generation of a residual positive τ_{yy}. In symmetrical turbulence the RMS fluctuation v' is assumed symmetrical about $\bar{v} = 0$. Considering unit volume of fluid, the upward bursting and downward sweeping masses ($\rho/2$, where ρ is the fluid density) and amounts of fluid momentum are equal. No residual τ_{yy} exists. In asymmetrical turbulence, on the other hand, the fluctuations are asymmetrical about $\bar{v} = 0$. In order to conserve momentum, the faster upward-moving bursting fluid must, on average, be of smaller mass than the slower downward-moving sweeping fluid. There exists an asymmetry, a, of the fluid masses involved in turbulence. Although momentum is conserved, there exists a net momentum flux, τ_{yy}, away from the bed whose magnitude is controlled by a.

continued on p. 139

Further information on Bagnold's dynamic theory of suspension
[*continued*]

masses and momenta involved in turbulence are equal. No residual stress of the form $|\tau_{yy}|$ can exist. In asymmetrical turbulence a positive skewness of measured vertical velocities may exist about the time-mean $\bar{v}$ of zero. In order to conserve momentum, the faster upward-moving fluid must, on average, be of smaller mass than the slower downward-

Fig. B6.3 'Bagnold tube' designed to test the idea of asymmetric turbulence. Water enters from the base and is distributed upwards in individual tubes to a baseplate (black circle). Here the water enters the top of the tube through a grid of small-diameter holes. The tube is sealed at the top and the water leaves the side connection via an array of larger holes in the baseplate. The net upward momentum flux that results from this artifice in the upper chamber causes a piece of wire gauze (white disc) to rise once the water supply is increased to a critical level.

continued on p. 140

moving fluid. There must exist a net momentum flux away from the bed whose magnitude is controlled by the mass asymmetry. [In the tube analogue of Fig. B6.3 a constant discharge of water now enters through smaller holes and leaves through bigger holes.] In symbols:

$$\tau_{yy} = \rho \int_0^{v_{max}} v^2(pv)\,dv - \rho \int_{v_{min}}^0 v^2(pv)\,dv \qquad (B6.20)$$

Adopting these ideas to modern turbulent structural models, we must assign the fast-moving upward masses to burst motions and the slow-moving downward masses to sweep motions. The theory is remarkably well supported by very accurate modern measurements of vertical turbulent velocities, which show very clear turbulent asymmetry in the area of the buffer layer. Additionally, the postulated τ_{yy} should be in equilibrium with an equal and opposite downthrust exerted by the suspended load by virtue of its immersed weight. That is:

$$\tau_{yy} = m_s' g \qquad (B6.21)$$

where m_s' is the immersed mass of suspended load. Experimental data gained from flume experiments bear out this postulate reasonably well (Leeder, 1983).

Further information on Bagnold's dynamic theory of suspension [*continued*]

---

sediment concentration, C_a, in the bedload zone (the 'reference level' problem).

6.10 A warning: nonequilibrium effects may dominate natural sediment transport systems

The transport of sediment out of catchment 'factories' has been described dynamically in the previous sections in terms of sediment transport by fluid stresses caused by the slip of turbulent water down gravity slopes. The horizontal tensor τ_{yx} transports bedload and the vertically acting τ_{yy} supports the suspended load. This is R.A. Bagnold's great legacy to fluvial hydraulics. Yet, comfortable as this is, mechanistic notions of equilibrium are ill-suited to studies of natural sedimentation. We need to introduce briefly the concept of water and sediment supply 'demons' that upset these notions and throw us back into disequilibrium. Such demons exist because of natural lags arising from river-independent supply (hillslope sediment runoff, rainsplash effects, etc.) and the inability of a turbulent flow to react instantaneously to oversupply, particularly of silt-sized and smaller particles. The

effect is well illustrated by measurements taken after desert flash floods by Laronne and Reid (1993). The existence of such demons should not depress us, however, since their recognition opens up a rich area of research concerned with quantifying all the natural inputs of sediment into the scheme of things.

6.11 Steady state, deposition or erosion: the sediment continuity equation

Perhaps the most fundamental question in sedimentology is the innocent-sounding 'How does one predict when sediment grains will be deposited or eroded?' It is only possible to answer this difficult question with a great deal of fudging. The starting point is an adaptation of the accounting system developed previously in the context of pure fluid flow itself (Chapter 4)—the continuity equation.

We have seen that fluid particles may accelerate or decelerate in response to changes at a point or over time along some distance. If the fluid in question is charged with sediment grains and the transporting machine is in a state of equilibrium, then any local deposition or erosion will cancel out and the system is in a steady state. Should the fluid be forced to change its velocity, say by a channel widening or contracting or a wind accelerating up the lee side of an obstacle, then the motive power available will change and deposition or erosion will result. In the simplest possible way we can thus make a box model of the accounting system:

deposition or erosion = (sediment in) minus (sediment out)

Further information on the derivation of continuity

All erosional or depositional systems in sedimentology or geomorphology must obey the law of mass conservation. For any system the difference between material coming into a control volume and that leaving the volume must represent either addition, subtraction or steady state of mass in the volume. Addition equates with deposition, and subtraction with erosion in our case. Material coming in may come downslope by local gravitational forces or be introduced by other independent sources. Material fluxes may vary with both time and space. How can we specify these fundamental concepts further? Following Wilson and Kirkby (1975) consider any land (or seafloor) surface over which sediment is transported from upslope to downslope across an element CD. Let i be the quantity of sediment transported in the x-direction. Let q be the amount of sediment added from outside the system, say by wind deposition or by contour currents. Then i and q can vary with both incremental space, δx, and time, δt. Let the surface have an elevation, y, which also varies with horizontal distance, x and time, t. The definition diagram in Fig. B6.4 sketches a concave surface that has grown by

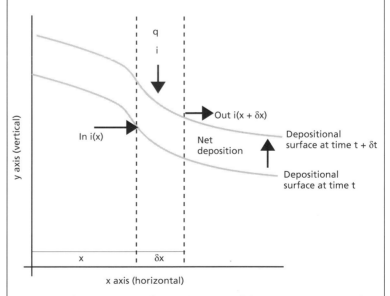

Fig. B6.4 Definition diagram for the derivation of the continuity equation for erosion, deposition and sediment transport (after Wilson & Kirkby, 1975).

continued on p. 142

deposition in time δt, defining an elemental area ABCD. For this area:

$$(\text{mass flux in}) - (\text{mass flux out}) = \text{increase of mass in ABCD} \quad \text{(B6.22)}$$

In the notation of the definition diagram:

$$(i_x - i_{x+\delta x})\delta t + q\delta x\delta t = \delta z\delta x \quad \text{(B6.23)}$$

Dividing by $\delta x\delta t$ gives

$$\frac{i_x - i_{x+\delta x}}{\delta x} + q = \frac{\delta z}{\delta t} \quad \text{(B6.24)}$$

Now, letting $\delta x\delta t$ tend to zero to obtain the exact partial differential equation that has universal validity:

$$-\frac{\partial i}{\partial x} + q = \frac{\partial z}{\partial t} \quad \text{(B6.25)}$$

We see that the rate of erosion or deposition (RHS) depends on the sign of the input term, q, and of the horizontal rate of change of the transport term, i. Note that the equation assumes no internal density changes, whereas in fact the natural system may vary in porosity due to deposition and compaction, and also no internal sinks or sources such as mineral dissolution or precipitation.

Further information on the derivation of continuity
[*continued*]

Further reading

Clifford *et al.* (1993) and Ashworth *et al.* (1996) are research monographs with many relevant papers on sediment transport. Thorne *et al.* (1987) deals with the research problems of sediment transport in gravel-bed rivers. In the absence of any modern readable text solely dedicated to the principles of sediment transport, a return to Bagnold's 1954(b) classic (for wind) and to his 1966(b) (for water) should provide the necessary inspiration. In this context it is worth the reader looking into the selected edition of Bagnold's writings published by the American Society of Civil Engineers (Thorne *et al.* 1988).

Part 4
Sediment Transport and
Sedimentary Structures

7 Bedforms and structures formed by unidirectional water flows over granular sediment

What a strange, strange boy
He sees the cars as sets of waves
Sequences of mass and space

Joni Mitchell, 'A Strange Boy', *Hejira*, Asylum Records

7.1 The 'trinity' of flow, transport and bedform

In previous chapters it was assumed, for simplicity, that sediment transport and deposition occur over flat beds. If this were truly the case, our subject would be an uninteresting one. Thankfully the real world contains a plethora of interesting forms and shapes that develop on beds of sediment during the transport of sediment grains. These bedforms occur on a variety of scales, from a few grain diameters to mountains of moving sand in deserts. The migration of bedforms leaves behind tell-tale structures in the deposited bed, enabling sedimentologists to read the rocks in a way that can often appear magical to other geoscientists.

Leaving aside the details of bedform morphology and structures for the moment, it is obvious that

Fig. 7.1 An illustration of the feedback between turbulent flow, sediment transport and bedform development; defining the 'trinity' of cause and effect relevant to bedform stability, sediment transport and flow modification. (After Leeder, 1983; Best, 1993.)

the fluid flow field will be greatly changed over and around any bedform as compared to flow over a flat bed. Complex interactions and feedbacks will thus occur (Fig. 7.1).

7.2 Current ripples

Current ripples are stable bedforms above the threshold for sediment movement on fine sand beds at relatively low flow strengths. They do not form in sands coarser than about 0.7 mm. On artificially smoothed beds, the forms evolve gradually from chance bed defects formed by impacting groups of turbulent sweep motions (Williams & Kemp, 1971; Best, 1992a; Fig. 7.2).

They may also form from initial bed irregularities well below this smooth bed threshold. Current ripples are asymmetric in the *xy* plane (parallel to flow), having a gentle sloping upstream or stoss side, sometimes with a prominent crestal platform, and a more steeply sloping (30°–35°) downstream or lee side (see Figs 7.3 & 7.4). Ripple height *h* ranges up to 0.04 m, ripple

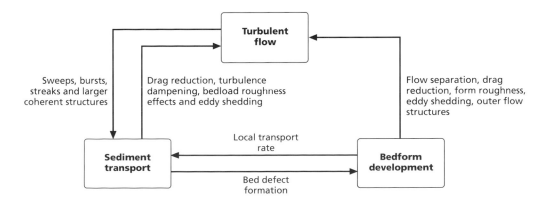

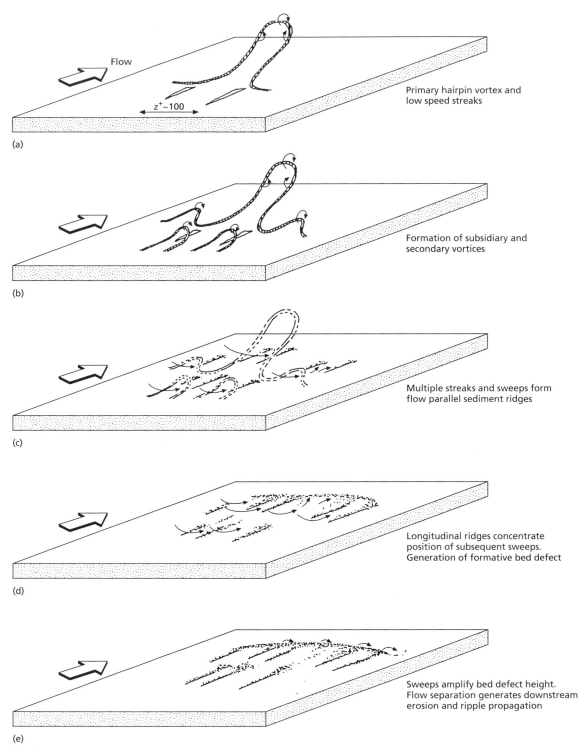

Flow

$z^+ \sim 100$

(a)

Primary hairpin vortex and
low speed streaks

(b)

Formation of subsidiary and
secondary vortices

(c)

Multiple streaks and sweeps form
flow parallel sediment ridges

(d)

Longitudinal ridges concentrate
position of subsequent sweeps.
Generation of formative bed defect

(e)

Sweeps amplify bed defect height.
Flow separation generates downstream
erosion and ripple propagation

Fig. 7.2 3D representations of the interaction between
primary and secondary hairpin vortices of turbulence and a
bed composed of granular solids. The interactions lead to
the production of transverse bed defect ridges, which flow
separation/reattachment processes amplify to ripples
by enhancing turbulent stresses downstream. (After
Best, 1992a.)

(a)

(b)

Fig. 7.3 (a) Sinuous and (subordinate) linguoid current ripples; flow from right to left; scale bar = 0.15 m; Solway Firth, Scotland. (b) Linguoid current ripples; flow from bottom to top; cigarette packet = 0.1 m long; Severn Estuary, England. *Continued on p. 148.*

wavelength λ to 0.5 m, with typical ripple indices (λ/h) of 10–40. Populations of current ripples show a variation of size and it is likely that individuals are continually forming, growing and decaying. There is no clear relationship between ripple size and either flow strength or water depth. However, λ does vary with grain size, being roughly λ = 1000d, although the plot shows much scatter. There is a clear and unequivocal separation of size (but not general form) between current ripples and the larger dune bedforms (Fig. 7.3).

Viewed from above, ripple crestlines may be straight, sinuous or linguoid (tongue-shaped) (Figs 7.3 & 7.5). It was previously thought that this increasing

(c)

(d)

Fig. 7.3 (*continued*) (c) Large-scale trough cross-stratification; flow towards observer; hammer shaft = 0.30 m; Old Red Sandstone, Lower Devonian, Welsh Borders. (d) Downward-dipping sets of large-scale cross-stratification; scale bar = 0.1 m; Fell Sandstone, Lower Carboniferous, Northumberland, England.

complexity occurred as flow velocity increased for a given depth. Careful experiments reveal that the straight- and sinuous-crested forms are metastable, always changing to linguoid ripples given sufficient time (Baas *et al.*, 1993; Baas, 1994). This equilibrium time is an inverse function of flow strength, reaching values of many hours or days for flows very close to the movement threshold. It may be objected that the experimental flows are all in laboratory *channels* and that wall effects inevitably give rise to secondary flows (Chapter 5) and hence to longitudinal, flow-parallel, elements in *any* bedform. Some support for this view comes from observations on unrestricted tidal flats where long transverse-crested forms are quite common.

On cohesive or other hard substrates (such as the floor of experimental flumes) and sediment beds where sand supply is limited, the disequilibrium current ripple forms take the shape of trains and rows

of isolated half-moons with elongated arms extending downcurrent. These lunate forms are termed *barchanoid*, after the equivalent larger-scale forms described from deserts (see Chapter 8).

Mapping of skin-friction lines from plaster-of-Paris models (a technique invented by D. Hopkins of Reading University) shows a pattern of flow separation at ripple crests with flow reattachment slightly downstream of the ripple trough (Figs 7.4 & 7.6; Allen, 1982). A captive bubble of recirculating fluid is thus held in the ripple lee, the size of which is a function of mean flow velocity and ripple height. Grains are moved in bedload up the ripple stoss side until they fall or diffuse from the separating flow at the crest to accumulate high up on the steep ripple lee face. Periodically the accumulated grains become unstable as the angle of accumulation exceeds the angle of initial yield. Small grain avalanches then occur, terminating at the toe of the lee face, and a single avalanche lamina is thus accreted on to the ripple lee. Ripple advance by lee slope deposition results in the flow attachment point shifting up the back of the downstream ripple where increased erosion occurs because of the very high turbulent stresses generated at the reattachment point. In this way the ripples

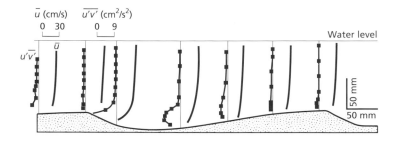

Fig. 7.4 Profiles of mean velocity (*u*) and a measure of turbulent intensity ($-\rho \overline{u'_{rms} v'_{rms}}$) as measured over a fixed and sand-coated experimental ripple model. Note the indication of a roller vortex in the ripple lee and the large amount of turbulence produced in the flow reattachment region. (Data of Sheen, 1964, as reported in Raudkivi, 1976.)

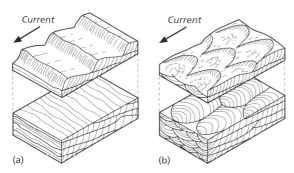

Fig. 7.5 Diagrams to show that the migration of (a) straight-crested and (b) curved-crested ripple and dune bedforms produces planar and trough cross-sets, respectively. (After Allen, 1970.) Note that preservation of successive cross-lamina sets necessarily involves a degree of net deposition. Most sets therefore climb at some angle from the local bed inclination (see Fig. 7.7).

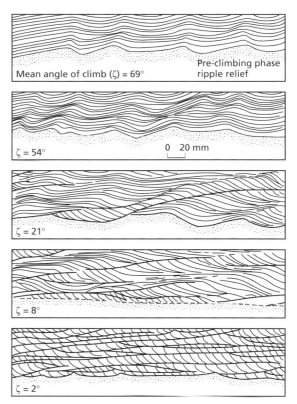

Fig. 7.7 Experimentally produced climbing-ripple cross-lamination seen in vertical profile parallel with flow. The increasing angle of climb from bottom to top is caused by the increasing rate of net vertical deposition relative to the speed of advance of the ripples. (After Allen, 1972.)

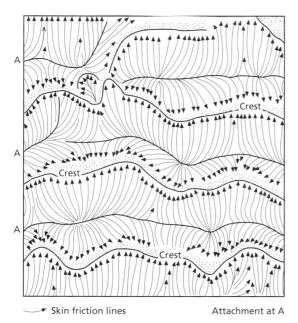

Skin friction lines   Attachment at A

Fig. 7.6 Pattern of skin-friction lines for flow (mean velocity 0.22 m/s, depth 0.095 m) over a current-rippled bed. Flow from bottom to top. The ripple crests mark lines of flow separation. The steep ripple avalanche faces are stippled. (After Allen, 1969.)

constantly shift downstream, preserving their overall equilibrium shapes, at a fairly constant velocity dependent upon the magnitude of the flow strength.

Sections cut through current ripples in the xy plane parallel to flow reveal successive accreted avalanche laminae (see Chapter 11 for details of such grain flows), which define the sedimentary structure known as small-scale cross-lamination. Sectioned normal to flow in the xz plane, the cross-laminae may be parallel and horizontal, defining planar cross-lamination, or trough-shaped, defining trough cross-lamination (Fig. 7.5). Tabular cross-laminae result from the migration of straight-crested ripples, whilst trough cross-laminae result as sinuous to linguoid ripples migrate forward into heel-shaped scour troughs eroded by the isolated separation eddies (Fig. 7.5). If there is no net sediment deposition at a particular point, then no cross-lamination can be produced other than that found within the individual ripple elements. When net deposition occurs, then a particular ripple crest will have a vertical component of motion as well as a horizontal component (Fig. 7.7). Sets of cross-lamination may thus be formed, bounded by erosive surfaces. The thickness of the sets is directly proportional to the rate of upward movement. Internally the set boundaries are seen to 'climb' at an angle to the horizontal (Fig. 7.7), and the structure is known as climbing-ripple cross-lamination. High angles of climb, with preservation of stoss-side laminae, indicate high rates of net deposition as in decelerating flows such as river floods or turbidity currents.

Close to the limits of their stability as flow strength increases, current ripples in very fine sands or silts give way, via 'washed-out' forms having similar wavelengths but lower heights, to plane bed conditions (see following sections). Characteristic scour-like features accompany the change in fine silty sediments (Van Gelder *et al.*, 1994). Ripples in fine to medium sands give way to dunes, but the nature of this transition is poorly understood.

7.3 Lower-stage plane beds and cluster bedforms

As noted above, ripples do not form in coarse sands ($d > 0.7$ mm); instead an equilibrium plane bed exists. In well-sorted sands this so-called lower-stage plane bed exhibits shallow scours and narrow irregular grooves 2–3 grain diameters deep over its surface. Net deposition on a lower-stage plane bed should give rise to crude planar laminations, but convincing examples have yet to be described from the sedimentary record.

As briefly noted above, increased sediment size leads to the production of transitional and then rough flow boundaries as the grains create their own near-bed flow field. This is accompanied by the increasing effects of eddies shed by the larger grains, chiefly lee-side eddies that are forms of Kelvin–Helmholtz waves, which periodically erupt into the free flow (Acarlar & Smith, 1987). In a poorly sorted bed composed of a variety of sizes, the coarsest ($d > 84\%$) cause lee-side sheltering effects that reduce drag and lift (Brayshaw *et al.*, 1983) and encourage grains to settle in the lee; stationary cluster bedforms thus arise on the bed.

7.4 Dunoids (bars, 2D dunes)

As the flow strength is slowly increased over transitional to rough beds, the lower-stage plane bed with its various defects and clusters gives way to irregular sheet-like to triangular-shaped bedforms termed two-dimensional dunes (Fig. 7.8), bars or, for coarser and poorly sorted sediments, bedload sheets (Whiting *et al.*, 1988; Bennett & Bridge, 1995). Because of the argument about the status of these forms and the plethora of names, I prefer to call them all *dunoids*. The higher forms have certain similarities to the general shapes of current ripples and dunes, but they generally have highly variable wavelengths and a large wavelength/height ratio. Dunoids closely resemble kinematic shock waves in their behaviour, the higher forms travelling faster and overtaking the smaller forms causing amalgamations (kinematic shock waves). Lee-side flow separation is poorly developed and so dunoid crestlines tend to be relatively straight, unbroken by isolated scour pools. The ineffective nature of separation is highlighted by the low form drag associated with dunoids. Internally the structures are structureless or characterized by the presence of weakly developed cross-stratification separated by more prominent horizontal to near-horizontal scoured surfaces.

7.5 Dunes

With increasing flow strength over sands and gravels, current ripples, lower-stage plane beds and the enigmatic dunoids give way to dunes. These large bedforms (Fig. 7.8) are similar to current ripples in general shape but are dynamically distinct and not simply large ripples as implied by the loose term 'megaripples'. This distinctiveness is indicated by:
- the lack of overlap between ripple and dune form indices;
- the positive relationship between dune dimensions and flow depth (Fig. 7.9);
- the fact that dunes do not form in sediment finer than about 0.1 mm;
- the observation that current ripples may be superimposed upon the backs of dunes (Fig. 7.10a) in an apparently equilibrium relationship;
- the strong interaction between lee-side flow vortices (Plate 3) and the upper flow.

Dune wavelengths commonly range from 0.6 m to hundreds of metres, and heights vary from 0.05 to 10.0 m or more. In plan view dune crests may be straight to strongly curved, but it is not known whether the former are equilibrium forms (cf. current ripples in section 7.2). The large size of many dunes makes study of their equilibrium shapes difficult; it is never clear how exposed dunes in tidal and river beds have been modified by the waning flows that led to

(a)

(b)

Fig. 7.8 (a) A straight-crested dune or bar with superimposed linguoid ripples formed during the falling stage of tidal ebb flow. (b) Sinuous dunes with well-developed scour pools. Solway Firth, Scotland.

their exposure. Detailed echo-sounding studies during bankfull discharges in rivers (Gabel, 1993) show that, like ripples, dunes undergo continuous birth, growth and decay, often with strong interactions between neighbours, including cannibalism and splitting.

The flow pattern over dunes is similar to that over ripples, with well-developed flow separation and reattachment. In addition large-scale advected eddy motions (boils, kolks) rich in suspended sediment occur. These are caused by shear instabilities and vortex interactions between the separating flow of the

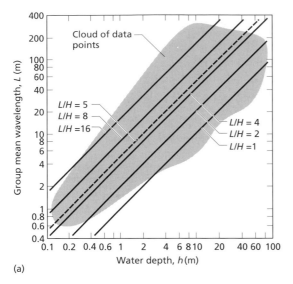

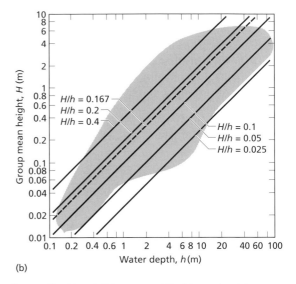

(a)

(b)

Fig. 7.9 Graphs to show the correlations between (a) dune group mean wavelength and mean water depth and (b) dune group mean height and mean water depth, both for one-way river flow only. (Data assembled by Allen, 1982.) Despite the considerable scatter of data (generalized here), it is apparent that the scale of dunes increases with boundary-layer thickness, implying, but not proving, that dune bedforms owe their origins to flow processes that affect the whole outer boundary layer.

free stream and the lee-side fluid closer to the bed (Muller & Gyr, 1982; Kostaschuk & Church, 1993; Bennett & Best, 1995, 1996; Plate 3; Fig. 7.11). These large eddy motions are not related to conventional bursting events rising from the buffer layer (they are far too large for that), but rather owe their origins to a process of periodic eddy shedding along the free shear layer at and downstream from the dune crest.

Dune migration gives rise to large-scale cross-stratification of planar or trough type in a similar way to that outlined for current ripples. Tangential contacts between the individual cross-laminae and the bounding set surface are encouraged by relatively weak lee separation eddies and by a high fallout rate of particles from suspension in the dune lee. Counterflow ripples (Boersma, 1967) result when grains are swept back up the lower parts of tangential foresets by near-bed flow in the separation bubble.

The regularly dipping cross-sets may also often be cut by erosive surfaces resulting from erosion of the dune crest and lee by falling-stage or low-stage flows (Jones & McCabe, 1980). Such features are preserved as reactivation surfaces (Collinson, 1970) within the migrating dune when normal avalanching events begin once more at rising and high flow stages. Sometimes, successive reactivation surfaces may enclose smaller-scale cross-stratification in an arrangement aptly known as downward-dipping cross-stratification. Here, small dune forms have migrated up the stoss side of the parent bedform *and* down the gently dipping lee side, where they are preserved as cross-sets (Banks, 1973).

Finally we must note that a major new development in dune bedform studies has come with detailed and documented confirmation from the Fraser River (Kostaschuk & Villard, 1996) of a state of dune development (previously reported by Smith & McLean,

Fig. 7.10 (*opposite*) (a) View over a sinuous dune crest to show scour pool and ripple fan on back of next dune downstream; scale = 0.5 m; Loughor Estuary, Swansea, Wales. (b) Primary current lineations in fine sands. (From Allen, 1964.) (c) Upper-phase plane beds; knife = 0.15 m long; St Bees Sandstone, Cumberland. (d) Train of antidunes (wavelength ≈ 0.3 m) in fast shallow tidal channel flow; Barmouth Estuary, Wales. (e) Train of upstream-breaking antidunes in tidal channel; flow left to right; shovel handle = 20 cm long; Solway Firth, Scotland.

(a)

(b)

(c)

(d)

(e)

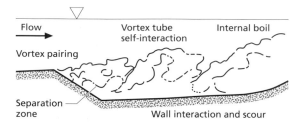

Fig. 7.11 Vortex development and 'boil' generation associated with the free shear layer of the dune flow separation zone. (After Muller & Gyr, 1986.)

1977) characterized by streamwise symmetry, lack of avalanche face and absence of lee-side flow separation. These forms may be transitional to upper-stage plane beds and are similar in general form to the 'humpback dunes' of Saunderson and Lockett (1983). At the time of writing, we know nothing of the internal structures of these symmetrical duneforms.

7.6 Upper-stage plane beds

As flow strength is further increased over fine to coarse sands, dunes give way, via characteristic humpback dunes (Saunderson & Lockett, 1983), to flattened forms (see previous section). Eventually, intense sediment transport occurs over an apparently featureless surface—this is the upper-stage plane bed (Fig. 7.10b,c). In fact, as careful experimentation with sensitive echo-soundings has made clear (Best & Bridge, 1992), the bed surface actually comprises small-amplitude/long-wavelength bedwaves. These bedwaves are asymmetric in the xy plane, range from 0.75 to 11 mm in height, from 0.7 to 1.0 m in wavelength and travel with velocities of up to 10 mm/s. Flow measurements suggest that the flow accelerates over the bedform crests and decelerates over the bedform troughs. Upon these bedwaves are superimposed flow-parallel ridges and hollows known as primary current lineation. Primary current lineation is the direct result of the viscous sublayer structure discussed in Chapter 5. Incoming sweeps, spaced parallel to flow, push grains aside to form the tiny grain ridges separated by broad troughs. It is important to emphasize that primary current lineation is not restricted to the upper-stage plane bed regime but may occur on the stoss side of ripples and dunes.

The low-relief bedwaves on upper-stage plane beds give rise to an internal structure of planar to wavy

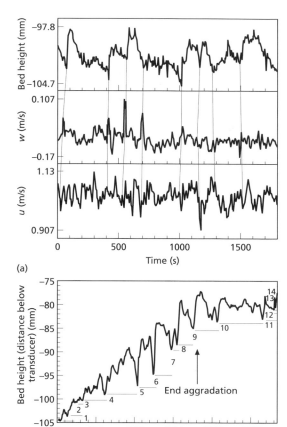

Fig. 7.12 (a) Elevation and velocity profiles over the bed during the passage of low-amplitude bedwaves on an upper-phase plane bed. (After Best & Bridge, 1992.) Note the correlations (dotted) of bedwave troughs with high spanwise (w) and low streamwise (u) velocities, indicating downstream decelerating and laterally accelerating flow conditions there. (b) Plot to show the aggradation of an experimental upper-phase plane bed with time. (After Best & Bridge, 1992.) The peaks and troughs originate as successive low-amplitude bedwaves pass over the aggrading bed. Numbers 1–12 indicate the laminae preserved in the final deposit that was subsequently cored and sliced. It can be seen that these laminae originate as erosively based features whose preservation potential is marked by the relative rates of bed aggradation, bedform height and wavelength, and bedform migration rate. (For details see Paola & Borgman, 1991; Bridge & Best, 1997.)

laminations that range between 5 and 20 grain diameters thick (Fig. 7.12). The mechanism of formation is revealed by high-resolution bed profiling and subsequent box-coring of aggrading experimental plane beds (Best & Bridge, 1992). Lamina preservation and thickness are dependent upon aggradation rate and the passage of bedwaves of varying size. Laminae are made visible by sediment sorting in the lee side of the waves, usually by infiltration of finer grains into an underlying surface. Ancient examples of upper-stage plane beds do not show preservation of the bedwaves themselves; rather, primary current lineations on the more-or-less planar stoss surfaces are the most obvious feature.

7.7 Antidunes, transverse ribs, chutes and pools, and related forms

Antidunes are sinusoidal forms with accompanying in-phase water waves that occur in supercritical flows (Fig. 7.13). Antidunes commonly occur in long trains; the individual waveforms may be stationary

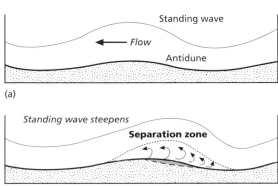

(a)

(b)

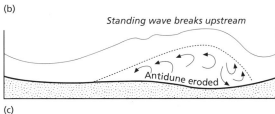

(c)

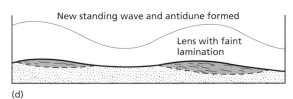

(d)

Fig. 7.13 Upstream breaking of a standing wave over an antidune bedform to produce faint upstream-dipping laminations as the form shifts upstream. (After Middleton, 1965.)

or undergo cycles of amplification and destruction in which they periodically steepen, move upstream and break up in a great rush of turbulence, the process then beginning again. The upstream migration gives rise to upstream-dipping sets of low-angle (< 10°) laminations (Middleton, 1965). The laminae are indistinct since the rapid upstream antidune migration is not accompanied by avalanching and grain sorting. The laminae have a low preservation potential since they are reworked during the cycle of renewed antidune growth; also any deceleration of the flow will destroy the laminations. Antidune bedforms formed on coarse, gravelly substrates have been termed transverse ribs (Koster, 1978), being low flow-transverse ridges a few grain diameters high. They are not uncommonly preserved on bar tops in braided stream settings. Spectacular large-scale gravel antidunes have been reported from the channel of the Burdekin River, Queensland (Alexander & Fielding, 1997). These flow-transverse bedforms had long wave-lengths (up to 19 m) and heights (up to 1 m), with steeper stoss (upstream) than lee slopes. Internally they featured erosive-based lenses of sandy gravel with low-angle *downstream-dipping* laminations whose long axis pebble fabric imbricated normally, i.e. dipping upstream.

Further increase in Froude number leads to formation of chute-and-pool bedforms. The chutes form in shallow, supercritical parts of a flow with high slopes, which end abruptly in deeper pools where the flow is subcritical. The upstream boundary of a pool is marked by rapid flow deceleration, the violently breaking water defining a hydraulic jump. Sediment accumulation may occur in the relatively tranquil pool region where steeply dipping backset laminations develop. The only known occurrence of preserved chute-and-pool laminations is in volcanic base surge deposits around the rims of maar-type volcanic craters (Schminke *et al.*, 1975).

7.8 Bedforms and sediment transport in poorly sorted sediment

Thus far we have discussed bedforms and sediment transport in moderately to well-sorted sediment. However, many natural sediments are poorly sorted and commonly comprise mixtures of sand- and gravel-sized grains. At low flow strengths it is common for the finer sand-grain fraction to be moulded into ripples, dunes or dunoids and for these familiar bedforms then to migrate over a more-or-less station-ary substrate of immobile coarse grains that are then left behind as a lag deposit layer, sometimes called an armoured layer (because the coarse grains protect the underlying bed like a layer of armour). As flow strength increases, the coarser grains are transported with the finer fraction at values of bed shear stress that may be much lower than those required to transport beds of similar-sized coarse grains alone. This is because the gravel grains in their matrix of finer sedi-ment protrude more from the bed; it is elementary to deduce that they are acted upon by greater fluid drag and lift forces and hence may be transported with and over the fine sediment. Also the coarser grains present obstacles to the flow when they are set in a matrix of finer grains; their own form drag creates a flow field that tends to scour the finer sediment around them, thus serving to mobilize them at values of bed shear stress well below those necessary to move coarser grains in beds of themselves. These properties cause bedforms to arise in poorly sorted sediments, particu-larly gravel–sand mixtures. The interaction of rolling grains causes particles to group, slow down and trap finer sediments in their interstices. Once a critical concentration of finer grains is reached, the friction between the large grains is reduced and they become mobile once more; thus the bed soon evolves into a pattern of flow-transverse alternations of finer and coarser sediment (the 'bedload sheets' of Whiting *et al.*, 1988).

In addition to the formation of flow-transverse fea-tures, it is quite common with poorly sorted sediment to find that the flow may organize the bed into flow-parallel ridges and troughs. Here the coarser sedi-ment comprises the ridges and the finer sediment the troughs. In such situations the size segregation is self-sustaining because the coarser ridges cause higher-than-average vertical turbulent motions, and vice versa for the finer troughs. A stable secondary circulation (the 'differential roughness secondary flow' of Pantin *et al.*, 1985) is thus set up that transfers upwelling fluid over the crests to downwelling fluid in the troughs where the local bed shear stress is greatest due to the turbulence produced. In channels, this is superimposed on the general downstream secondary flow briefly dis-cussed in Chapter 4. The sediment size-sorting effect in flow-parallel ridges may thus *result* from Taylor–Görtler vortices or *cause* the formation of the vortices, depending upon the geometry of the flow situation.

7.9 Bedform phase diagrams

It is traditional to separate the bedforms discussed above into two broad groups (Simons *et al.*, 1965). Ripples, lower-phase plane beds and dunes define a lower flow regime where flow resistance is relatively high and where water waves and large eddy bursts at the water surface are out of phase with the bed undula-tions. Upper plane beds, antidunes and chute-and-pool structures define an upper flow regime where flow resistance is relatively low and where surface waves, where present, are in phase with any bed undulations. The previously discussed discovery of low-amplitude bedwaves on upper-stage plane beds that are out of phase with the overlying diverging and converging flow throws this traditional classification into doubt.

Friction coefficients calculated for current ripples and dunes are 2–5 times those calculated for upper- and lower-stage plane beds. The increase is due to turbulent energy created and destroyed by lee-side flow separation and reattachment. The extra fric-tion is termed *form drag* and should be distinguished from the surface drag due to the sediment grains themselves. As noted in Chapter 5, form and surface drag combine to determine the overall roughness of any bed.

This contrast in friction coefficients raises an important point concerning the use of bedform phase diagrams in which bed shear stress or flow power is used as the measure of flow strength. The applied fluid shear, τ, may be written as:

$$\tau = \frac{\rho f \bar{u}^2}{8} \tag{7.1}$$

where f is the Darcy–Weisbach friction coefficient, ρ is fluid density and $\bar{u}$ is mean flow velocity. We see that τ is a direct function of the friction coefficient. Since the friction coefficient is itself dependent on the type of bedform developed, we can see that bed shear must itself be a function of bedform type (Southard, 1971). We could therefore imagine the situation where the same shear stress could be produced by slower flow over a rough (e.g. dune) boundary or a faster flow over a smooth (e.g. upper plane bed) flow boundary. This effect is partly responsible for the overlap of the dune and upper-phase plane bed fields in plots in-volving bed shear stress. This 'overlap problem' may be overcome by plotting bedform phase diagrams as mean flow velocity vs. grain size (Fig. 7.14).

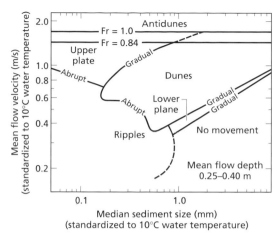

Fig. 7.14 The simplest and most easily interpreted bedform phase diagram, involving a plot of mean flow velocity vs. mean sediment size for a wide range of flow depths. Variables are standardized to water temperature, viscosity and density conditions at 10 °C. (Data of Southard & Boguchwal, 1990a, as slightly modified by Ashley, 1990.)

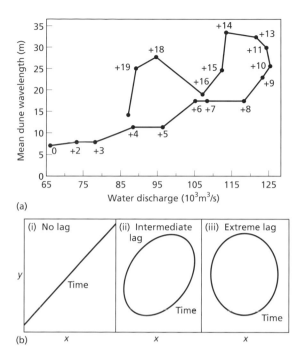

Fig. 7.15 (a) Variation of mean dune wavelength with water discharge over a period of 19 days in the Fraser River, British Columbia, Canada. Note the well-developed lag effect. (b) General form of the variation of a quantity, y (such as the dune wavelength), as a function of another quantity, x (such as river discharge), when both are also functions of time. (After Allen, 1973.)

A final word of caution concerning the application of laboratory-determined bedform states in velocity–shear stress–grain size space to natural environments (where water depth and outer flow thickness are very much greater) comes from the recent discovery of natural symmetrical dunes under upper-stage plane bed conditions (Kostaschuk & Villard, 1996). The implications of these observations to the use of bedform phase diagrams have yet to be analysed.

7.10 Bedform 'lag' effects

Many of the remarks made so far in this chapter assume the presence of an 'equilibrium' bed state adjusted to a steady flow. However, natural flows of water and wind are unsteady on a variety of scales, e.g. tidal flows for hours, and river and air flows for weeks or months. Subaqueous dunes that form in response to a steady flow may persist for a considerable time as the flow decays into the ripple stability field. Within the dune stability field, changes in water depth during rising or falling river stage may cause changes in the dune wavelength or height to lag behind the flow (Fig. 7.15). We may idealize our discussion with reference to Fig. 7.15(b), in which graphs of y and x are plotted so that both variables are also functions of time; y could be dune wavelength

whilst x could be discharge. If no lag exists, then a straight-line plot results. Lag reaches extreme values of 90°.

Bedform lag in natural environments means that we should be very careful in interpreting field measurements of bedform sizes in relation to existing flow conditions. Criteria for bedform equilibrium should be investigated. One relevant example might be the equations relating dune wavelength and height to water depth, in which very considerable scatter was evident. Much of this scatter could have resulted from lag effects.

7.11 Bedform theory

Ripples are stable at low values of applied fluid shear stress in fine sediment. The spanwise sequence of fluid bursts and sweeps in the viscous sublayer plays an important role here (Williams & Kemp, 1971; Best,

1992a; Fig. 7.2). Entrainment of grains during a sweep is followed by deposition as the sweep fluid decelerates. All over a sand bed, made artificially plane prior to experiment, tiny discontinuities are formed from heaps of deposited grains 2–3 diameters high. Certain heaps now begin to influence flow structure as flow separation and reattachment occur on their lee sides. These heaps become magnified and propagate downstream since the turbulent stresses at attachment erode too much material to be held in transport. More heaps are thus formed downstream, which in turn give rise to flow separation, and so on. Tiny current ripples now form, coalesce and interact all over the bed. An equilibrium ripple assemblage evolves after an hour or so. The ripples grow to an equilibrium height and wavelength determined by the size of the lee-side separation eddy. They are not related to outer flow structures since they have no relationship of magnitude to flow depth. The curved crest equilibrium form suggests some control by longitudinal vortices.

The well-known inability of coarser sands and gravels of grain size $\gtrsim 0.6$ mm to form ripple-like bedforms has been explained by changes brought about in the flow by such coarse bed grains. It can be no coincidence that the threshold for motion of coarse sands of diameter ~0.6 mm is the point at which the viscous sublayer begins to be broken and disrupted by the grain roughness. It is thus expected that at this point the near-bed boundary layer over small bed defects produced by turbulent streaks should be more intensely mixed due to the production of eddies shed off from the lee sides of both bed grains and those in saltation transport. This increase of mixing is seen in the increased values of the vertical component of flow turbulent stresses (Grass, 1971; Ligrani & Moffat, 1986; Kirkbride, 1993). The effect of enhanced mixing is to steepen the velocity gradient and decrease the pressure rise at the bed in the lee of defects. This causes separation to be inhibited and the defects are unable to amplify to form ripples. An alternative view is that the 0.6 mm limit represents the onset of grain saltations whose wavelength matches that of the bed defects produced by turbulence. Thus grains saltate their way out of the lee-side separation eddies and defects are unable to amplify.

The positive relationship between dune height, wavelength and flow depth indicates that the magnitude of dunes is somehow related to thickness of the boundary layer or flow depth. In view of the prominent occurrence of large-scale turbulent eddies that are generated in the separation-zone free shear layer, it is likely that dunes grow to an equilibrium determined by the effects of such large-scale eddies on the transport of suspended sediment. The instability of a ripple or lower plane bed field as flow strength is increased may be due to the random growth of very large ripples or dunoids (Fig. 7.16) that are able to trigger large-scale eddy formation in their downstream free shear layers. These statistical freaks (rogue ripples) somehow trigger the appearance of like forms downstream and so a dune field gradually equilibrates (see Bennett & Best, 1996).

The change from dune to upper plane bed states occurs at large values of applied bed shear stress. It is thought that the change may somehow be due to the effects of increasing suspended load and bedload sediment transport upon the structure of the turbulent flow close to the bed. It is possible that turbulence may be suppressed by increasing concentrations of sediment in transport and hence the likelihood that turbulent flow separation will occur becomes less. This reasoning is supported by the fact that the dune stability field is much reduced or absent in sediments below 0.1 mm diameter (Baas et al., 1993; van der Berg & van Gelder, 1993), and also by the discovery of natural symmetrical duneforms without lee-side flow separation at high bed shear stresses (Kostaschuk & Villard, 1996).

Antidunes occur as stable forms when the flow Froude number is > 0.84, approximately indicative of rapid (supercritical) flow, and are thus common in fast, shallow free-surface flows. They may also occur in two-layer flows. Antidune wavelength of free-surface flows is related to the square of the mean flow velocity (Kennedy, 1963) by:

$$\lambda = \frac{2\pi u^2}{g} \tag{7.2}$$

The corresponding equation for two-layer flows is a little more complex and will not be discussed here (for a good discussion see Prave, 1990). The rare occurrence of antidunes preserved on bedding planes facilitates direct determination of palaeoflow velocity if λ can be measured.

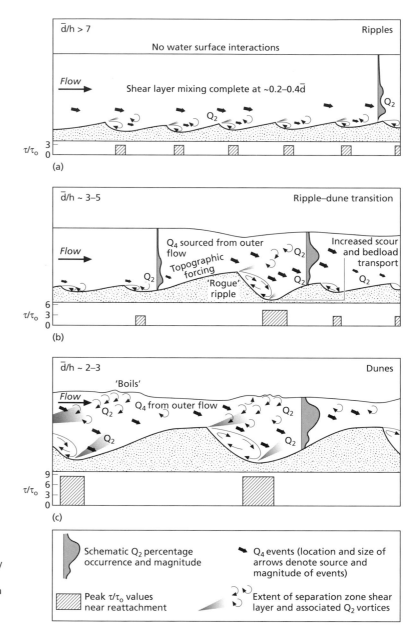

Fig. 7.16 Sketches of bedform changes across the ripple-to-dune transition, involving the growth of 'rogue' current ripples of larger-than-average size that trigger changes in the response of the sediment bed and turbulence, chiefly involving the increased involvement of the outer flow and the production of large Kelvin–Helmholtz eddies. (After Bennett & Best, 1996.)

7.12 Measurement of palaeocurrents and problems arising from trough-shaped sets of cross-stratification

It is commonplace for students (and professionals) to misunderstand the very principles of palaeocurrent measurement in the field or in core by an insufficient appreciation of the influence of the 3D shape of cross-bedding. It is absolutely necessary to establish the 3D geometry of any set before a palaeocurrent measurement is used to reconstruct palaeoflow (DeCelles *et al.*, 1983; Fig. 7.17). This is because the existence

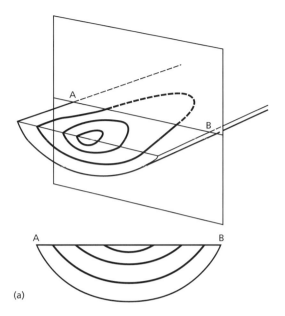

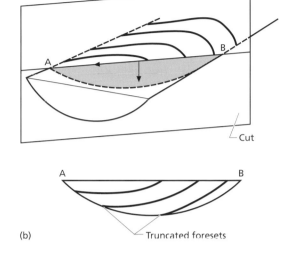

Fig. 7.17 (a) Transverse-vertical and (b) oblique-vertical cuts through cylindrical trough cross-stratification to illustrate the problems involved in the measurement of palaeocurrents from trough-shaped cross-stratification. (After DeCelles *et al.*, 1983.)

of the simplest tabular cross-sets, when the angle of maximum dip may be directly measured or reconstructed from the stereographic projection of two apparent dips, must be firmly established from field observations in well-exposed 3D sections. Trough-shaped sets cannot be reconstructed from two meas-

urements of apparent dip. They must be measured when the trough shape may be seen in plan view, as a measurement along the trough axis, or from numerous measurements around a trough as assembled as a stereographic projection.

Further reading

Allen (1982) is a fundamental, and weighty, reference for *all* bedforms and sedimentary structures discussed in this and subsequent chapters. Collinson and Thompson (1982) is a lower-level source for many examples.

8 Bedforms and structures formed by atmospheric flows

I shall walk out beyond Walmer Castle, by way of the sand-dunes: there is a wilderness of time in that arenaceous world.

Patrick O'Brian [Stephen Maturin, to his diary], *Post Captain*, HarperCollins

8.1 Introduction: some contrasts between air and water flows

In order to understand aeolian (from Aeolius, King of the winds) sediment transport and bedform development, it is necessary to appreciate a number of critical differences between air and water as transport agents. These arise because of differences in fluid material properties, magnitude of stresses exerted and thickness scale of the boundary layers (Table 8.1). Note the following points:

1 The low shearing stresses set up by moving air mean that the range of grain sizes transported through bedforms is very much more restricted than for water (essentially sand to granule grades).

2 The low buoyancy force experienced by silicate grains in air means that conditions at the bed during bedload transport are dominated by collision effects as saltating grains exchange momentum with the bed.

3 Significant 'splashup' effects occur each time a grain collides with a bed of like grains, leading to a chain reaction downwind.

4 Splashdown causes a significant 'creep' component of grains that are nudged along by the successive impacts, a mode of transport termed reptation.

5 Energetic collisions mean that wind-blown transport is very effective in abrading and rounding both sediment grains and impact surfaces.

6 The bedload layer of rebounding grains is much thicker in air and its effect is to add significant roughness to the flow boundary.

7 Suspension transport of sand-sized and larger grains by fluid turbulence is much more difficult in air than in water because of the small buoyancy forces acting.

Table 8.1 Some physical contrasts between air and water flows.

	Air	Water
Density (kg/m³)	1.3	1000
Quartz/fluid density ratio	2039	2.65
Buoyant force per unit volume	12.8	9810
Viscosity (N s/m²)	1.78×10^{-5}	1.00×10^{-3}
Stokes velocity, U_s, for 1 mm diameter sand (m/s)	~8	~0.15
Representative bed shear stresses exerted (N/m²)	~0.088 for a wind speed of about 6 m/s measured 1 m above the bed (shear velocity $u_* = 0.26$ m/s)	~68 for a comparable shear velocity
Critical shear velocity (u_*) needed for motion of 0.5 mm sand grain (m/s)	0.35	0.02

8 The inability of air to suspend coarser sand grains and the high fall velocity of those finer sands that do become suspended means that sorting processes in the relatively sheltered lee side of dune-like bedforms are dominated by periodic grainflows on the higher slip-faces and by more continuous rapid grain settling over the whole lee side, but particularly in the lower areas; this combination leads to a regular alternation of distinctive laminations.

9 The turbulent microstructure of the moving air plays little role in the initiation of ripple-sized bedforms.

10 The atmospheric boundary layer (10^2–10^4 m) is thick compared to even the deepest river or tidal shelf, so that the outer flow must play a role in determining bedform scale, notwithstanding major differences in near-bed mechanics.

11 The rapid response of air masses to heating and cooling causes rapid daily vertical variations in boundary-layer density (density may also be affected by evaporation). Important short-term buoyancy forces are set up, which cause many atmospheric flows to depart from the condition of neutral stability that

Further information on bedload transport by wind

More momentum is extracted by grains from unit volume of moving air than in equivalent water since each transported grain must be accelerated in the wind stream. This momentum is constantly given up to the bed upon grain collision, the transfer process between grain and fluid causing increased drag on the flowing air and a marked inflection of any measured velocity profile. The horizontal fluid velocity in and above the bedload layer is reduced compared to that of a sediment-free flow (Fig. B8.1) since the sediment grains exert a much greater force on the wind. The fluid shear stress is negligible at the bed during saltation transport, increasing with height as the corresponding solid stress exerted by the grains decreases away from the bed. At all times the sum of fluid and solid stresses is constant. The various feedback processes acting in the wind boundary layer are illustrated in Fig. B8.2.

It is easy to calculate the drag exerted on a moving bedload of sand in air. One sand grain of mass m starts at zero velocity and acquires a forward velocity U from the wind over the length L of its travel. So, momentum extracted from wind per unit length is mU/L, and thus a mass q_s of sand in saltation moving over unit width per unit time will lose q_sU/L momentum per second per unit surface area. Momentum loss is a stress and is a measure of the resistance offered to the air by the moving grains. Thus writing τ as the shear stress and rearranging for q_s we have $q_s = \tau U/L$. Since shear stress is proportional to shear velocity squared, we have a nonlinear relation for the transport rate of sand in air. Bagnold originally developed this approach (see more recent accounts by McEwan & Willetts, 1994), illustrating the nonlinearity by calculating that a strong wind blowing at 16 m/s (measured 1 m above the ground) moves as much sand in one day as is moved in three weeks by a wind blowing steadily at 8 m/s. The rate of bedload transport also varies strongly according to the slope of the natural surface over which the transport occurs (Hardisty & Whitehouse, 1988). Thus on upwind dune flanks the transport rate is less and on lee-side slopes the transport is greater than that predicted by the rate equations derived above. This is because the impact of saltating grains on a slope causes reptating grains to fall back or move forward extra increments according to the slope direction.

continued on p. 163

Further information on bedload transport by wind [*continued*]

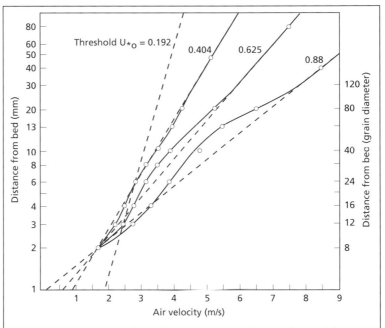

Fig. B8.1 Graph to show (1) the effective roughness height of 3 mm (about 12 grain diameters) for winds of various strengths transporting fine sand as saltating bedload, and (2) the deceleration of the near-surface wind due to momentum extraction by the saltating grains in bedload transport extending progressively outwards into the boundary layer (to a maximum of about 100 grain diameters) as the shear velocity is progressively increased. (After Bagnold, 1936; 1973.)

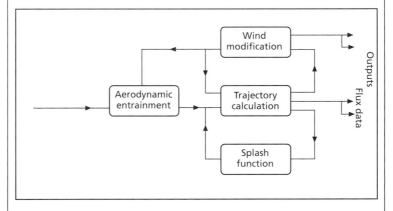

Fig. B8.2 Flow diagram for numerically modelling the chief processes and interactions responsible for wind-blown saltation transport. The simulation begins with a clean-air logarithmic wind profile and, after initial entrainment of grains by direct fluid forces, the saltation is increasingly dominated by grain–bed collisions and ballistic splashups. The bedload reaches equilibrium when the mass of grains in transit causes the wind close to the bed to decelerate below the fluid threshold for motion. (After McEwan & Willetts, 1993.)

(a)

(b)

(c)

(d)

Fig. 8.1 (a) Ballistic ripples illustrating grain-size control of wavelength; the smaller ripples on the left are in finer-grained sand and have been affected by a later gentle wind, at 90° to the first, which did not disturb the coarser sand on the right. (From Wilson, 1973.) (b) Aerial view of aklé dunes (scale unknown) from Utah. (From Cooke & Warren, 1973.) (c) Barchan dune advancing across a lag gravel pavement; La Joya, South Peru. (From Cooke & Warren, 1973.) (d) Draa comprising superimposed aklé dunes; draa height 30 m; Erg Occidental, Algeria. (Photo courtesy of Ian Davidson.) *Continued opposite*.

dominate freshwater flows (Wyngaard, 1992). Little work has been undertaken into the effects of convective buoyancy on near-surface wind vectors, sediment transport and bedform development. In the discussions that follow it is usually assumed that stable boundary-layer flow exists, i.e. no buoyancy effects exist.

8.2 Aeolian bedforms in general

Bedforms under atmospheric flows range over more than four orders of magnitude in size, from the lowly ripples familiar from the action of summer breezes on dry beach sand, to the gigantic hills of sand captured by aerial and satellite images over deserts (Fig. 8.1). In his landmark works Wilson (1972a,b) defined three distinct bedform groups from the Algerian Erg Oriental by plotting grain size against bedform wavelength. In ascending order of magnitude, these are wind (ballistic) ripples and ridges, dunes and draas. Subsequent studies (Wasson & Hyde, 1983) indicate strong overlap between the proposed dune and draa fields, negating the Wilson concept. Recent work makes it clear that bedform scale and type depend upon both the availability of sand and the nature of the wind regime. The term *draa* may still be applied to large-scale composite bedforms made up of a hierarchy of smaller duneforms of various shapes.

(e)

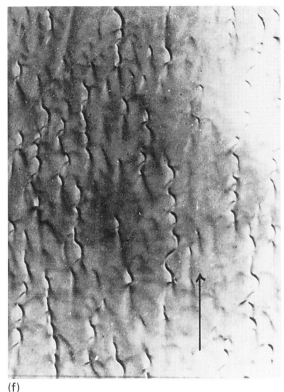

(f)

Fig. 8.1 (*continued*) (e) Aerial view of barchanoid draa dunes; Sahara, Algeria. (After Wilson, 1972a.) (f) Aerial view of 'meandering' seif dunes migrating across a lag of coarse sands; Edeycnubari, Libya. (After Wilson, 1972a.)

8.3 Ballistic ripples and ridges

Ballistic ripples and ridges form a continuous series (Fig. 8.2) with wavelengths 0.02–2.0 m and heights from a few millimetres to 0.1 m. Ripple indices generally fall between 8 and 50.

Wavelength increases linearly with increasing coarseness of the mean grain size (Fig. 8.3) or with decreased sorting. For the same grain size and sorting, wavelength increases with increasing wind strength. The smaller ripples have slightly sinuous crests that are very persistent normal to the wind flow direction. More strongly curved-crested forms, analogous to subaqueous linguoid ripples, sometimes with superimposed smaller forms (Seppälä & Linde, 1978), occur in faster wind flows blowing very fine to fine sands. In flow-wise cross-section the forms are asymmetric, with a gently sloping, slightly convex lee side and more steeply dipping (usually around 20°) concave stoss side. A common attribute of all but the smallest forms in very well-sorted sands is the concentration of coarser grains in crestal areas.

There is often no clear internal structure of cross-laminations in aeolian ripples, unlike their subaqueous counterparts. This is because the ripples migrate less by repeated lee-side avalanches than by saltation bombardment. A particularly common feature of strata deposited by migrating ballistic ripples is a strong reverse grading or bimodality of grain size (Hunter, 1977)—a consequence of the concentration of coarser creepload along the ripple crests (Anderson & Bunas, 1993). The bounding surfaces between the ripple laminae frequently dip slightly upwind (Hunter, 1977), indicating a slow build-up of the sediment bed during ripple migration, a feature analogous to bedset boundaries in subaqueous climbing-ripple cross-laminations (Fig. 8.4).

Concerning the origin of aeolian ripples, the superficial resemblance of form to subaqueous ripples

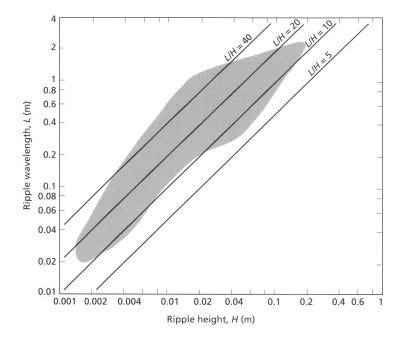

Fig. 8.2 Graph to show correlation between wavelength and height of aeolian ballistic ripples. Shaded area indicates envelope of data points. (After Allen, 1982, and sources cited therein.)

Fig. 8.3 Normal ballistic ripples in fine sand passing laterally from a dune margin into long-wavelength granule ridges developed at the margin of a fluvial channel. Great Sand Dunes, Colorado, USA.

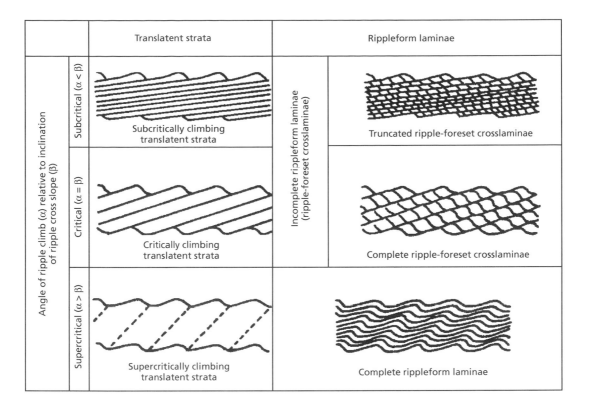

Fig. 8.4 Types of small-scale (sets are centimetre–decimetre in scale) structures produced by aeolian climbing ripples in areas of sand suffering net accumulation with time. (After Hunter, 1977.) The dipping set boundaries are marked by finer silts, and sometimes clays, trapped and overrun in ripple troughs; they may appear as 'pin-stripe' laminae (Fryberger & Schenk, 1988) when differentially cemented.

hides a fundamental contrast in the mechanism of formation. As discussed in the introduction to this chapter, saltating grains in air are much 'heavier' than saltating grains in water because of the greatly increased density ratio of solid to fluid. Conditions on the air–bed interface are thus dominated by grain splashdown effects (Bagnold, 1954b; Anderson & Bunas, 1993) and not by the small-scale viscous sublayer streaks that seem to control grain movement and ripple initiation over flow-produced defects on the water–bed interface (see Chapter 7). In air, fine sand grains move by saltation, and upon impact their kinetic energy is such that they are capable of nudging coarser grains along (up to six times their diameter) by intermittent rolling or sliding (termed reptation by some authors) to form the part (up to 25%) of the total sediment transported as creepload. Ripples produced by this ballistic process will increase in wavelength with increased grain size and decreased grain sorting since the saltation jump length of fine particles increases markedly with the size of the bed grains with which they collide. In addition, the impact of high-energy saltating grains is found to eject the smaller grains preferentially. This leaves zones of coarser grains that migrate in reptation (small hops or excursions about a grain diameter or so at a time) as patches over downwind plinths of finer grains protected in the lee of the patch. Intense bombardment on the upwind side of the patches thus alternates with lee-side sheltering and a stable ripple form gradually emerges with a characteristic coarser crestal grain size (Anderson & Bunas, 1993).

Despite the dynamic contrasts between wind and water, the process of ripple formation and the final ripple forms are sufficiently similar for us to conclude that some universal law is responsible. This law operates on small moving bed defects to make them larger and more regular. Recent developments (Anderson & Bunas, 1993; Werner, 1995) concerns the recognition that aeolian sediment transport and bedform development from small initial or random defects represents a complex system from which regular forms arise (or emerge) due to the convergence of a state known as an attractor. Bedform development from an initial random or flat bed is thus an example of self-organizing behaviour. A simple simulation model for sediment transport generates dunes of shape and size that depend upon the nonlinear transport law we briefly derived at the beginning of this chapter and upon the increased probability of transport over a nonerodible substrate (Werner, 1995). The model has not been applied to erodible substrates, but doubtless a similar behaviour is to be expected.

8.4 Dunes in general

Aeolian dunes show more diverse morphologies than do their subaqueous counterparts. A simple division (Fig. 8.5) into longitudinal (streamwise), trans-

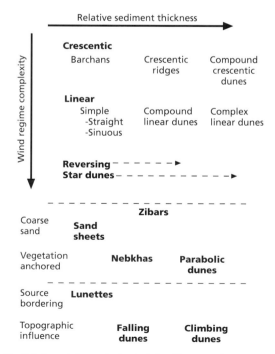

Fig. 8.5 A useful morphological classification of desert dunes. (After Lancaster, 1995, and sources cited therein.)

verse (spanwise) and complex forms with respect to the local mean resultant wind direction is useful, although transitional forms occur and many dunes commonly show combinations of all elements. A second control is exerted by the abundance of sand supply. Vegetation may play an important role in stabilizing and modifying duneforms (see Livingstone & Thomas, 1993). There are also a variety of dune types related to the topographic forcing of the wind by isolated hills and escarpments.

8.5 Flow-transverse dunes

Flow-transverse dunes occur where the predominant seasonal winds of importance for sand transport are unidirectional. There is a continuum of flow-transverse forms related to the availability of sand cover (Fig. 8.6). Crescent-shaped barchan dunes, with their 'wings' tapering downwind, are perhaps the most evocative and widely known flow-transverse dune type, even though they are a minority in most deserts. The barchanoid form is not restricted, however, to desert dunes; we have briefly discussed previously (Chapter 7) that current ripples take on such characteristic forms in water flows over an immovable or cohesive bed with insufficient sand supply to mould complete ripples. Similarly, barchan dunes occur in areas of reduced sand supply, with individual dunes separated from their neighbours by either solid rock floor or by immobile coarse pebbles (the *desert reg*, a lag deposit). Barchans and other transverse forms occur because, after sand is collected in an initial patch, continued vertical and downwind growth of the patch leads to a convex, mound-like form, a morphology that causes flow separation to occur and the production of a slipface (Bagnold, 1954b; Kocurek *et al.*, 1992; Fig. 8.7). Barchan wings form because sediment transport rates are higher on the margins of a patch of sand where intergranular frictional effects during saltation impact are much reduced and rebound is more efficient.

Successive barchans may propagate downwind from these wings. Barchan dunes are typically 1–20 m high, 25–250 m wavelength and up to 300 m wide, with migration rates of up to 30 m/yr. The migration saga of 'Bagnold's barchan' over 57 years at a mean rate of around 7.5 m/yr makes interesting reading (Haynes, 1989), not least because it is evident that such forms may migrate very far from their ultimate sand source. The internal structure of a barchan is

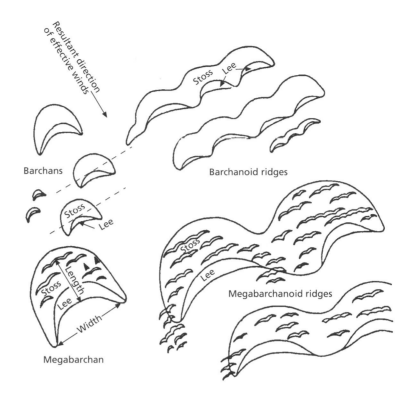

Fig. 8.6 Sketches in *xz* space (plan view) to show the various sorts of flow-transverse dunes. Sand supply and availability increase from top to bottom. (After Breed, 1977.)

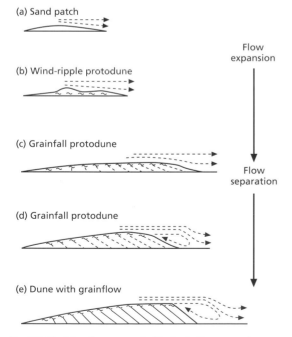

Fig. 8.7 Sketches in *xy* space to illustrate the successive stages of dune formation from an initial sand patch. Dune dimensions of metres high and tens of metres wavelength. (After Kocurek *et al.*, 1992.)

shown in Fig. 8.8. The rarity of isolated barchans in sand seas (ergs) with plentiful supply means that they are rarely likely to be preserved in the stratigraphic record. Isolated domal dunes without a prominent slip-face seem to form from the degradation of barchanoid dunes during long periods of gentle wind flows. As expected, such dunes show complex internal cross-stratification patterns.

Amongst the transverse dunes, sinuous-crested types, termed *aklé dunes* (Fig. 8.1b), are common in areas of plentiful sand supply. These show slipfaces orientated normal to the local flow vectors on dune lee sides, giving rise to large internal sets of cross-stratification. Frequent internal reactivation surfaces and downward-dipping cross-sets result from periodic modifications of dune shape by aberrant winds. Aklé dunes clearly indicate minimal effects of longitudinal secondary flow vortices. The origin of the simple aklé dune is still not adequately known. It is possible that an analogy with subaqueous dunes is apposite. This would require the wavelength of aeolian dunes (usually several hundreds of metres) to be related to boundary-layer thickness (again, usually several hundreds of metres), with dune wavelength controlled by the repeat distance of

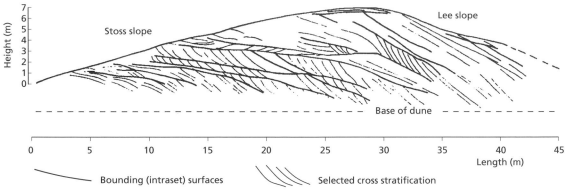

Fig. 8.8 Internal structure of a barchanoid dune from White Sands, New Mexico, USA. Note that the form has clearly evolved in morphology as it has grown, from a domal dune with internal down-dipping sets indicating passage of small dunes over it, to a much higher barchanoid form with an avalanche face occupying most of the steep lee slope. (After McKee, 1966.)

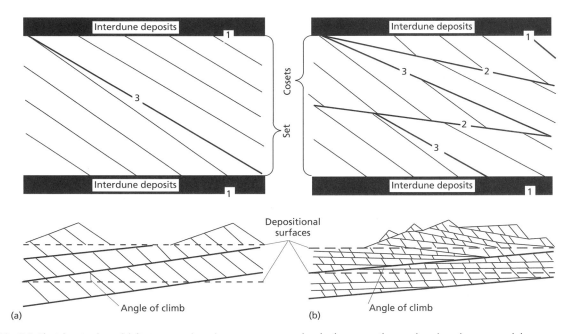

Fig. 8.9 Sketches to show (a) first-, second- and third-order bounding surfaces in dunes and (b) the migration and build-up of the depositional surface and the internal structures produced as it is traversed by simple dunes, on the one hand, and compound draas with superimposed migrating dunes, on the other. (After Kocurek, 1988.)

Fig. 8.10 Structures associated with the migration of small dunes over a larger flow-transverse draa-like form. Photo of trench section shows sets of aeolian cross-stratification with labelled prominent downwind-dipping bounding surfaces, Abu Dhabi, Trucial Coast. (From Bristow *et al.*, 1996; photo courtesy of C. Bristow.)

large-scale vortical events generated in the lee-side separation zone (see Chapter 7).

Draa bedforms (see Wilson, 1972a; McKee, 1978) are composite, usually transverse, elements with wavelengths up to 4000 m and heights up to 400 m. Draas are made up of superimposed dunes of all kinds (Fig. 8.1d). They may be aklé or barchanoid in plan. Draas take a long time to form, requiring appreciable original sand cover to provide an adequate original nucleus. Some draas show giant slipfaces up to 50 m high, but many do not and show instead dune migration both up the stoss side and down the lee side at fairly shallow angles. The resulting internal structure of a draa-like composite bedform features various orders of upwind- and downwind-dipping erosional and depositional surfaces (Figs 8.9 & 8.10). In ancient deposits, identification and mapping of these surfaces in three dimensions is needed to determine the morphology of the draa responsible (Kocurek, 1981; Fryberger, 1993; see also Chapter 16).

8.6 Flow-parallel dunes

Sharp-crested longitudinal dunes are often referred to as seif dunes (from the Arabic for sword), individual examples of which may sometimes be traced for many

tens of kilometres. Dune heights range up to 50 m or more, with typical lateral spacing between dunes of several hundred metres. Dune coalescence produces Y-shaped junctions, which most commonly fork upwind (Folk, 1971). Some seif dunes show a sinuous planform (Fig. 8.1f) or have periodic humps. The classic explanation for the latter is that they arise when barchan dunes are subjected to winds from two directions at acute angles to each other (Bagnold, 1954b; Tsoar, 1983; Wopfner & Twidale, 1988; Fig. 8.11). One barchan wind becomes elongated, later to become the nucleus of a new barchan as the wind re-establishes itself in its former mode. The resultant beaded seif dune (Bagnold, 1954b) has its long axis orientated parallel to the resultant of the two wind azimuths. This theory is broadly supported by flow visualization studies on the beaded seif dunes of the Sinai desert (Tsoar, 1983; Fig. 8.12), where the oblique incidence of seasonal winds to seif crestlines causes lee-side helical flow spirals to be set up. Internally, seif dunes show a bimodal pattern of large-scale cross-stratification produced by avalanche accretion of alternate sides of the dune during seasonal flows with high angles of incidence to the local crest direction (Bagnold, 1954; Tsoar, 1982; Fig. 8.13).

Concerning the origins of longitudinal aeolian dunes in general, it has been proposed that the presence of streamwise secondary flow is of major importance (Bagnold, 1954b; Wilson 1972a,b; Cooke & Warren, 1973). By analogy with the commonly observed windrows of blown snow or sand over immobile surfaces, longitudinal dunes may develop along the axis of the meeting point of pairs of oppositely rotating streamwise vortices. Finer saltating sands are thus

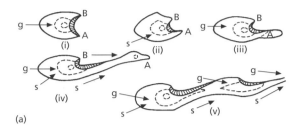

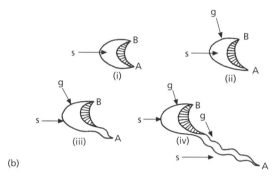

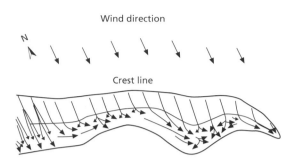

Fig. 8.12 The pattern of airflow (visualized by smoke canisters) close to the bed of an obliquely incident wind over a sinuous seif dune. (After Tsoar, 1978.)

Fig. 8.11 Bagnold's (1954b) classic model and a modification by Tsoar (1984) for the origin of beaded seif dunes from barchans by the alternation of bidirectional seasonal strong and gentle winds.

always swept inwards in broad lanes where deposition occurs and, given sufficient sand supply, the duneform grows into equilibrium with the flow. Once formed, the dunes will reinforce the secondary flow cells. Although an attractive theory, and despite many observations (e.g. Wopfner & Twidale, 1988), it has not yet been proven that small-scale sand windrows

may grow large-scale dunes. Opportunities for natural experiments are hindered by the very large scale of the effects searched for. A closer comparison in terms of scale might be made with linear cloud formations (cloud 'streets'), whose persistence and wavelength resemble linear dunes (Hanna, 1969). The much larger atmospheric secondary flows held by some to be responsible for linear dunes probably owe their origin to a combination of wind shear upon convecting air masses (Bagnold, 1954b; Hanna, 1969). But the role of convection currents on aeolian bedform development is poorly documented since Bagnold (1954b) originally speculated on the consequences of boundary-layer shear transforming such currents into longitudinal flow rolls in the immobile substrates between longitudinal dunes. There is thus still considerable disagreement over the origin and dynamics of even the simplest linear and transverse aeolian dunes (see Pye & Tsoar, 1990; Lancaster, 1995).

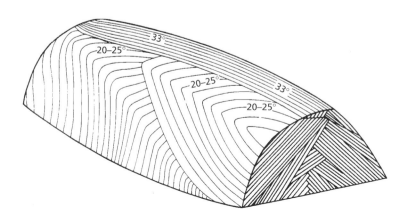

Fig. 8.13 The 3D internal structure of a typical seif dune. (After Tsoar, 1982.)

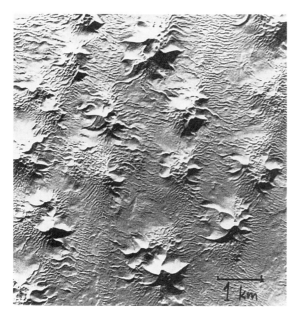

Fig. 8.14 Aerial photo of spectacular star-shaped dunes from the Erg Oriental, Algeria. (From Wilson, 1972a.)

8.7 Complex flow dunes

Spectacular star-shaped dunes (also known as rhourds; Fig. 8.14; Plate 4) commonly range from 500 to 1000 m wavelength and from 50 to 150 m height. The forms have central peaks about which curved crests radiate like vortex lines. They may be spaced randomly, separated by immobile rock or gravel substrates, or in rows, and seem to arise from the interaction of multidirectional regional winds (Lancaster, 1989) with, less certainly, local winds due to convected air masses. The flow over these forms is, not surprisingly, particularly complicated (Fig. 8.15), and the details of their internal structure remain poorly known. Clemmenson (1987) is the only person to have identified the cross-stratified deposits of an ancient star dune.

8.8 Vegetated parabolic dunes

These curious three-dimensional forms (see Pye, 1982, 1993; Figs 8.16 & 8.17) have the opposite orientation with respect to the unidirectional wind from barchans, i.e. the long tailing 'arms' of the dune point upwind.

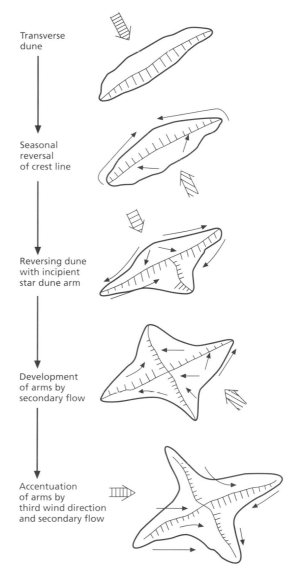

Fig. 8.15 Possible origin of star-shaped dunes by the development of secondary flow circulations as originally transverse dunes migrate into an area characterized by multidirectional winds. (After Lancaster, 1989.)

They owe their origins to the interaction between vegetation and an active source of sand, commonly point-sourced from a central 'blowout' where sand is liberated by intense turbulent shear. Vegetation growth on the relatively inactive 'arms' of the dune

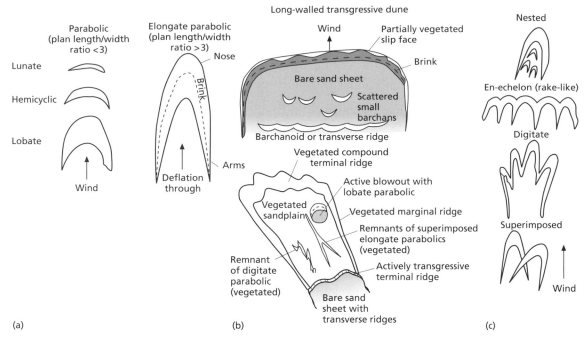

Fig. 8.16 (*above*) Types of (a) simple, (b) complex and (c) compound parabolic dunes. (After Pye, 1993.)

Fig. 8.17 (*below*) The supposed evolution of a parabolic dune from northern Queensland. (After Pye, 1982.)

Schematic representation of evolutionary development of elongate parabolic dunes at Cape Flattery

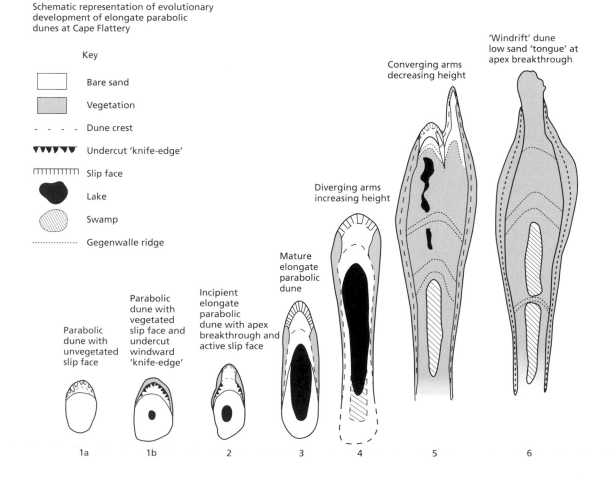

stabilizes these as ridges, and hence with time the point source continues to supply sand downwind to the active convexo-concave core of the duneform, causing the 'arms' to elongate markedly. In many areas, once the central source of sand and the active dune core become exhausted with fresh sand, all that is left to observe from aerial photographs are the long (sometimes tens of kilometres) flow-parallel 'arms', a situation that mimics the form of normal longitudinal dunes (and leads to sedimentary confusion amongst the unaware). Sometimes an elongate lake forms in the central part of the dune.

Further reading

Lancaster (1995) is the most up-to-date account of desert bedforms. Pye and Tsoar (1990) is also useful. Greeley and Iverson (1985) gives stimulating interplanetary perspectives. Read Bagnold (1954b) again for inspiration.

9 Oscillatory water waves, combined flows and tides: their bedforms and structures

At the Place for Pulling up Boats
(one word in Gaelic) the tide is full.
It seeps over the grass, stealthy as a robber.
Which it is.

Norman MacCaig, 'Two Thieves', *Collected Poems*, Chatto and Windus

9.1 Introduction

Waves are widespread, usually periodic, phenomena that transfer energy and sometimes mass. Thus physicists postulate the existence of sound, shock and electromagnetic waves, and we can observe waves of concentration each time we enter and leave a stationary or slowly moving traffic jam. A great range of waveforms and associated energies affect the atmosphere and oceans, with periods ranging from 10^{-2} to 10^5 s for the ocean waves (Fig. 9.1). Many waves are in motion, travelling from here to there as progressive waves, although some are of too low frequency to observe directly, like the tides. Yet others are standing waves, like the resonant standing oscillations present in many coastal inlets and estuaries.

The commonest visible manifestations of wave motion are the surface waveforms of lakes and oceans. In the oceans, waves are usually superimposed on a flowing tidal or storm current of greater or lesser strength, and these tidal and combined flows are particularly important in marine sediment transport. In later chapters we shall discuss waves at density interfaces in connection with the motion of density or turbidity currents and the astonishing solitary waves seen in Nature as tidal bores, tsunamis and reflected density currents.

Surface water waves are more-or-less regular periodic disturbances of the water surface created by surface shear due to blowing wind, the regular or irregular impact of objects, and the sudden motion of the adjacent land or bottom due to faulting. We can make waves of period 1–10 s or so in laboratory tanks using paddles or other oscillatory devices. The chief features of an oscillatory wave (Fig. 9.2) are:

- crest and trough;
- wavelength, λ;
- frequency, σ;

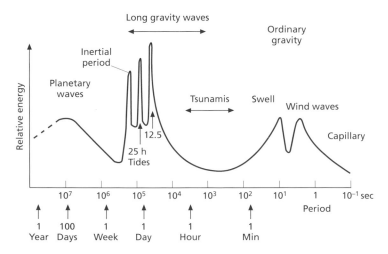

Fig. 9.1 The relative energy spectrum of various oceanic wave motions. (After Pond & Pickard, 1983.)

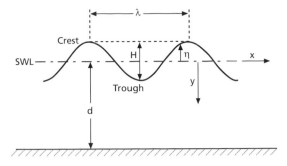

Fig. 9.2 Parameters of surface water waves.

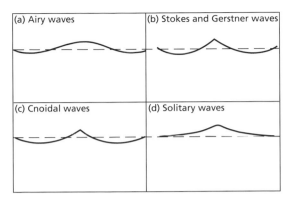

Fig. 9.3 Profiles of wave types derived by mathematical analysis. (a) Airy wave theory gives simple harmonic or sinusoidal forms and is applicable to waves of small amplitude in deep water. (b) Stokes and Gerstner waves are trochoidal in form and the theories have been applied to waves of finite amplitude in all water depths. (c) Cnoidal waves are of finite amplitude for intermediate to shallow water. (d) Solitary waves are single forms (more like heaps of water on an otherwise uniform surface) that move water mass forwards in shallow water.

- displacement of the water surface from its still water level, η;
- wave height, H;
- speed of passage, c, sometimes called celerity;
- period that one wavelength takes to travel, T;
- wavenumber, k, the number of wavelengths per unit distance.

From these definitions we can see that the following simple expressions apply:

$$\sigma = 1/T \quad \text{or} \quad \sigma = 2\pi/T \text{ (radian frequency)}$$
$$c = \lambda/T \quad \text{or} \quad c = \sigma/k$$
$$T = \lambda/c \tag{9.1}$$
$$\lambda = cT$$
$$k = 1/\lambda \quad \text{or} \quad k = 2\pi/T \text{ (radian frequency)}$$

The overall shape of a wave follows a curve-like form and we must use this often smoothly varying property as a simple mathematical guide to our study of wave physics. It is a common mistake to imagine all water waves as heaps and troughs of water moving along a surface. Just one waveform, solitary waves (see Chapter 11), and also waves travelling in shallow water, approximate to this model. In all other oscillatory waves, it is just the wave energy that is transferred, without involving net forward water motion. The stationary observer, fixing their gaze at a particular point such as a partially submerged marker post, will see the water surface rise and fall up the post as a wave passes by through one whole wavelength. This rise and fall signifies the conversion of wave potential to kinetic energy. Wind-generated waves of the sort we are now considering are a form of surface gravity wave driven by a balance between fluid inertia and the restoring force of gravity. The simplest approach is to set the shape of the waveform along an x–y graph

and consider that the periodic motion of y will be a function of distance x, wave height, wavelength and celerity. Thus we have $y = f(x,H,\lambda,c)$. Attempts to investigate wave motion in a more rigorous manner assume that the wave surface displacement may be approximated by curves of various shapes (Fig. 9.3), the simplest of which is a harmonic motion used in linear (Airy) wave theory.

9.2 Simple wave theory

In simple harmonic motion the water particle displacement, y, is given by:

$$y = H \sin \omega t \tag{9.2}$$

where H is wave height, t is time and ω is angular velocity (radians per second, rad/s). Since $\omega = 2\pi/T$, where T is the periodic time (wave period), then

$$y = H \sin\left(\frac{2\pi}{T}t\right) \tag{9.3}$$

or, since $T = \lambda/c$:

$$y = aH \sin\left(\frac{2\pi c}{\lambda}t\right) \tag{9.4}$$

Simple sinusoidal waves of small amplitude in deep water cause motions that cannot reach the bottom ('deep' is formally defined as water depth $> \lambda/2$). In this small-amplitude wave theory approach, the water is assumed to be inviscid and thus irrotational. The job of solving the equations of motion for this case is a little more involved than readers of this text need bother with—see below for the simplest approach and Lighthill (1978) or Acheson (1990) for more advanced but very accessible accounts. The result gives a simple expression for the wave speed as:

$$c = \sqrt{g\lambda/2\pi} = gT/2\pi \qquad (9.5)$$

The coefficient $\sqrt{g/2\pi}$ is constant and so for measurements made in SI units we have:

$$c = 1.25\sqrt{\lambda} \qquad (9.6)$$

and

$$T = 0.80\sqrt{\lambda} \qquad (9.7)$$

We see that surface gravity waves travelling over very deep water are dispersive in the sense that their rate of forward motion is directly dependent upon wavelength. Wave height and, of course, water depth (by definition) play no role in determining wave speed. The linear theory of sinusoidal waves on deep water predicts that at any fixed point the fluid speed caused by wave motion remains constant whilst the direction of motion rotates with angular velocity ω. Any fixed surface particle must undergo a circular rotation below deep-water waves (Fig. 9.4). The radius of these water orbitals, as they are called, decreases exponentially below the surface according to $H \exp(-2\pi z/L)$. But it should not be imagined that the orbital paths illustrated in Figs 9.4(a) and 9.5(a) are stationary rotating vortices. At any one instant the water in any vertical plane is all moving uniformly according to position with respect to the wavy surface (Fig. 9.4b). These instantaneous sheets of flow lie adjacent to other sheets slightly out of phase and so on. Most wave energy (about 95%) is concentrated in the half-wavelength or so depth below the mean water surface.

In Nature it is found that waves may be quite different in shape and dynamics from the simple linear theory of sinusoidal deep-water waves. One reason is that waves appear more irregular, being combinations of different-frequency sinusoidal waves,

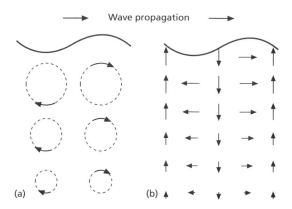

Fig. 9.4 Two views of orbital motions. Although every particle rotates about a time-mean circular motion, the circles drawn are *not* stationary vortices shearing past each other. The solid arrows show instantaneous motion vectors at each arrowhead. (From Vogel, 1994.)

all superimposed. These can be disentangled using energy–frequency plots and Fourier decomposition, the energy in a wave being proportional to the square of its height. A second reason is that the deep-water wave theory developed above fails when the ratio of water depth to wavelength falls below about 0.5. As deep-water waves pass into shallow water, defined formally as $d < \lambda/20$, they suffer attenuation through bottom friction and induce significant horizontal motions on the bottom. The waves take on new forms, with more-pointed crests and flatter troughs. Such waves are of great interest to sedimentologists. After a transitional period, when wave speed becomes increasingly affected by water depth, shallow-water gravity waves move with a velocity that is proportional to the square root of the water depth, independent of wavelength or period:

$$c = \sqrt{gh} \qquad (9.8)$$

The wave orbits are elliptical at all depths (Fig. 9.6), with increasing ellipticity towards the bottom, culminating at the bed as horizontal straight lines representing to-and-fro motion. Under such conditions the maximum orbital horizontal velocity is given by:

$$u_{max} = \frac{H}{2h}\sqrt{gh} \qquad (9.9)$$

where H is wave height.

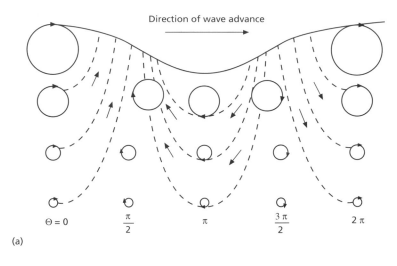

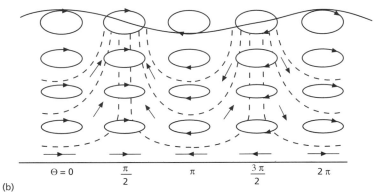

Fig. 9.5 Orbits, relative orbital velocities and streamlines for (a) deep-water waves and (b) shallow-water waves. In the latter case, although the orbits are shown closed, there is in fact a net transfer of water mass in the direction of wave propagation (see Figs 9.6 & 9.10). (After McLellan, 1965.)

Fig. 9.6 Flow visualization of suspended particles photographed orbiting under a shallow-water wave traversing one wavelength left to right. Wave amplitude is 0.04L and water depth is 0.22L. The clockwise orbits pass from near-circular at the surface to ellipses of increasing elongation towards the bottom. Some open loops indicate a slow near-surface drift to the right and a near-bed drift to the left. (From van Dyke, 1982.)

Take the water wave shown in Fig. B9.1(a) and in a thought experiment (you are a stationary observer, i.e. Eulerian frame applies) make the waveform stationary by imposing an opposite water flow of velocity $-c$. The speed of water, u, in the wave orbital motion interacts with the opposing flow to give a net motion $u-c$ at the wave crest and $-u-c$ in the trough. The speed u is the distance of the wave orbit, the circumference πH, travelled over the period T, i.e. $u = \pi H/T$. Note carefully that this orbital speed for water particles is very much less than the wave speed (see below). Recalling Bernoulli's theorem for the energy budget along a streamline in an inviscid fluid,

More information on water wave theory

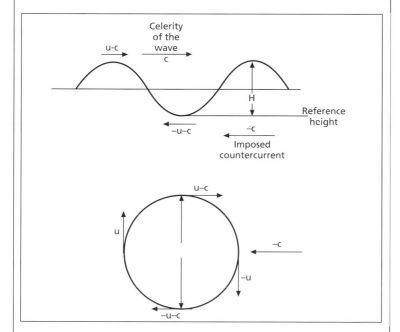

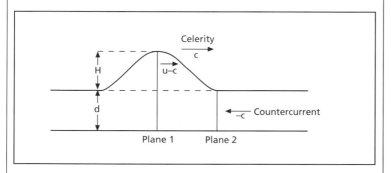

Fig. B9.1 Sketches to show the artifice of bringing (a) a progressive surface sinusoidal wave and (b) a solitary wave to a halt by imposing a countercurrent of velocity $-c$. (After Denny, 1993.)

continued on p. 181

More information on water wave theory [*continued*]

$$\frac{\rho u^2}{2} + p + \rho g y = \text{constant} \tag{B9.1}$$

and noting that the water wave surface is a streamline, we apply the theorem to a point at the crest and another point in the trough of any deep-water wave. Further, we neglect the tiny difference in pressure (atmospheric in this case) between trough and crest (but *not* in the water column beneath) and divide through by the density term to eliminate it. We then get the equality for energy along the surface streamline of wave and trough (with trough as reference level) as:

$$\frac{u_1^2}{2} + gH = \frac{u_2^2}{2} \tag{B9.2}$$

Making use of $u = \pi H/T$, $u_1 = u - c$ and $u_2 = -u - c$, some arithmetic, rearranging and cancelling gives:

$$c = \frac{gT}{2\pi} = \sqrt{\frac{g\lambda}{2\pi}} \tag{B9.3}$$

We can utilize a similar simple approach to determine the speed u of a shallow-water or solitary wave like a bore or tsunami (Fig. B9.1b). Here we place u_1 far upstream on the level water surface and u_2 on the wave crest. The speed of the water at point 1 is now $-c$ and at point 2, $u - c$. Inserting these values in eqn (B9.2) we get, again after arithmetic, rearranging and cancelling:

$$c^2 = u^2 - 2uc + c^2 + 2gH \tag{B9.4}$$

We can simplify this expression by cancelling the c^2 terms and neglecting the square of u here because $u \ll c$. To see this we can write the ratio of c to u as:

$$\frac{c}{u} = \frac{gT^2}{2\pi^2 H} = \frac{\lambda}{\pi H} \tag{B9.5}$$

and note that, for most waves, height is much less than wavelength (a supposition that breaks down for many true solitary waves!). We now have the simple expression:

$$c = \frac{gH}{u} \tag{B9.6}$$

To find u we make use of the continuity principle, that the discharge of fluid through any two sections of the flow depth must be equal. Thus at points 1 and 2, for unit widths normal to the section, we must have the products of flow speed and cross-sectional area as:

$$-ch = (u - c)(h + H) \tag{B9.7}$$

or

$$u = c - \frac{ch}{h + H} \tag{B9.8}$$

continued on p. 182

Substituting for u into eqn (B9.6) we have:

$$c = \sqrt{g(h + H)} \qquad\qquad (B9.9)$$

or, more simply still, when wave height is much less than water depth:

$$c \approx \sqrt{gh} \qquad\qquad (B9.10)$$

Full development of linear wave theory (with decreasing mathematical difficulty see Lamb, 1948; Lighthill, 1978; Acheson, 1993) enables us to obtain a continuous function for the influence of water depth on wave speed through the expression:

$$c = \sqrt{\frac{g\lambda}{2\pi}} \sqrt{\tan h \left(\frac{2\pi h}{\lambda} \right)} \qquad\qquad (B9.11)$$

The tanh (hyperbolic tangent) term approaches unity for deep-water waves and tends to $\sqrt{2\pi h/\lambda}$ for shallow water, the latter then giving the required expression for shallow-water waves derived already, viz. $c = \sqrt{gh}$.

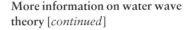

More information on water wave theory [*continued*]

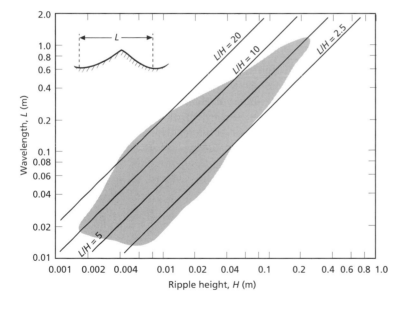

Fig. 9.7 Wavelength and height of wave ripples with an indication of the asymmetry provided by the ratio a/b (inset). Note that the strongly asymmetric ripples have higher ripple indices, i.e. they are higher relative to their wavelength, than the more symmetrical wave ripples, a trend carried on into true combined flow/wave ripples. Shaded area is an envelope of data points. (After Allen, 1982.)

9.3 Near-bed flow and bedforms

It follows from our discussion of particle paths in shallow water that to-and-fro bottom motions will cause shear stresses to be set up. The effect of these is to cause grain movement by rolling and saltation at some critical wave condition on an initially plane sand bottom. The formation of symmetrical (oscilla-tion) wave-formed ripples follows, whose crests are usually very persistent laterally but which bifurcate in a characteristic manner. The wave-formed ripples vary greatly in size since they are dependent only upon the dimensions of surface waves. They commonly vary in wavelength between 0.009 and 2.0 m and in height from 0.003 to 0.25 m, the ripple index varying between 4 and 13 (Fig. 9.7).

Fig. 9.8 Flow visualization of steady streaming above a ripple model. Scale bar = 1 cm. (After Honji *et al.*, 1980).

Let us consider the evolution of wave-formed ripples in more detail. Classic experiments with an underwater oscillating table (Bagnold, 1946) showed that, once the threshold for motion is reached, rolling and saltating grains tend to come repeatedly to rest along crests that lie normal to the oscillation direction. The grains at the ripple crests rest at the angle of residual yield. At low values of bed shear stress the ripple crests are low, perhaps < 20 grain diameters high, and sweep to and fro with broad, flat or gently curved troughs in which no grain movement occurs. (Note that such rolling-grain ripples have proved difficult to produce in other experimental set-ups in which the water is oscillated, rather than the bed as in Bagnold's experiments.) The detailed long-term average effects of many passing surface waves on the pattern of flow and sediment transport over a stable bed of 2D rolling-grain ripples of low amplitude is fascinating and beautiful. Close to the bed, fluid and sediment pass from trough to crest and back again in circulating loops. These were first visualized using ink by G.H. Darwin (son of Charles) in 1884 and termed by him 'ink mushrooms'. Reaching further from the bed and sourced in trough or crest depending upon wave stroke are more sluggish arcuate streamlines, termed 'ink trees' by Darwin (Fig. 9.8).

As bed shear stress is increased by increasing the wave speed or amplitude, the rolling-grain ripple crests reach a critical height, causing the formation of flow separation vortices on either side of a symmetrical ripple crest during the to-and-fro bottom motion of the water (Bagnold, 1946). The vortices are able to suspend finer sediment grains as they flip from side to side of the ripple crest during the passage of overlying waves, and it is the onset of this suspension that controls the development of the new symmetrical ripple forms, termed vortex ripples by Bagnold, but more widely known nowadays simply as oscillation or wave ripples. These are the common symmetric ripples seen on beaches.

Initially two-dimensional, as the oscillatory speed is increased, particularly at longer oscillation periods (2–20 s), wave ripples become increasingly three-dimensional (Southard *et al.*, 1990; Southard, 1991). For very fine to fine sands, ripple wavelengths on the sediment bed increase from a few centimetres for periods of < 2 s to greater than 1 m for periods > 10 s. For a given maximum orbital speed the ripple spacing increases with increasing wave period, the ratio of near-bed orbital diameter to ripple wavelength being, very approximately, one-half. The near-bed concentration of suspended sediment increases with orbital speed and is strongly dependent upon the vertical advection of sediment associated with the ejection of separation vortices from the lee of ripple crests (Fig. 9.9a). These vortices interact in unknown ways as they move away from the bed in the oscillatory turbulent boundary layer. Ultimately oscillation ripples are washed out and a plane bed with intense sediment transport is established.

The vortices set up on either side of developing wave ripple crests during passage of individual surface waves scour sand from the nascent ripple troughs and deposit it as laminations dipping concordant

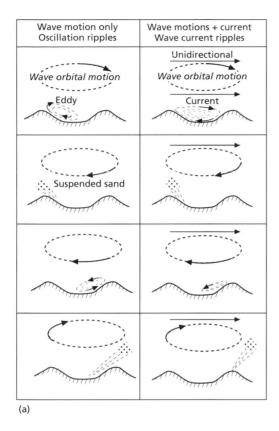

Wave motion only Oscillation ripples	Wave motions + current Wave current ripples

(a)

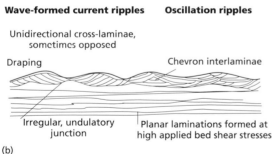

Wave-formed current ripples   **Oscillation ripples**

Unidirectional cross-laminae, sometimes opposed

Draping   Chevron interlaminae

Irregular, undulatory junction   Planar laminations formed at high applied bed shear stresses

(b)

Fig. 9.9 (a) The relationship between sand transport over a rippled sand bed and the orbital motions of wave action with and without a superimposed one-way current. (After Komar, 1976, from the original data of Inman & Bowen, 1963.) (b) Some characteristic internal features of wave-formed and combined wave–current ripples. (After de Raaf *et al.*, 1977.)

with, and on either side of, the symmetrical ripple crestal form. This increases ripple amplitude and thus decreases the ripple index or wavelength/height ratio. Sections through wave ripples (Fig. 9.9b) thus reveal

an internal structure of chevron-like laminae accreted on to either side of the stationary ripple crest as successive vortex motions are shed off from the ripple crest lee and the suspended sediment falls out on to the lee side below.

In water shallow relative to wave height, the approximations made by Airy wave theory break down as boundary-layer Reynolds stresses play an increasingly important role in wave mechanics (Longuet-Higgins, 1953). There is an increasingly important net landward transport of water (causing a 'radiation stress', see Chapter 25), seen from flow visualization photographs (Fig. 9.10) as successive wave orbitals that fail to close in the direction of propagation. It is important to stress that this net shoreward motion is not dependent upon any added tidal or storm-driven current (these *do* also occur, of course, see Chapter 25) and varies in magnitude according to the local conditions of wave type, flow depth, etc. At low magnitudes the effect causes more-or-less pristine symmetrical oscillation ripples to migrate very slowly in the direction of slow net forward motion. Should this motion be accompanied by net sedimentation, then the production of stacked sequences of translatory ripple form set laminae may occur (Fig. 9.11).

It is a very difficult matter to use inverse methods to estimate the possible magnitude of ancient waves from the observed dimensions of wave-formed ripples. Although the ratio of water orbit to ripple wavelength in shallow water is, very approximately, about one-half, this relationship does not hold for deeper-water waves and varies with grain size. Thus it is difficult to calculate ancient wave and water conditions (wave celerity, wavelength and height; water depth) unless there is independent evidence for water depth in addition to the commonly measurable quantities, ripple wavelength and ripple grain size, something that is rather rare in the sedimentary record. One may finally mention the problems that will arise because of the 'wave group' effect, whereby mean wave height may be a fairly meaningless parameter given the correlation of maximum sediment transport efficiency under the largest waves in any given group (see Osborne & Greenwood, 1993). Detailed accounts of protocols to establish ancient wave parameters, with illuminating case histories, are presented by Komar (1974), P.A. Allen (1984), Clifton and Dingler (1984) and Diem (1985).

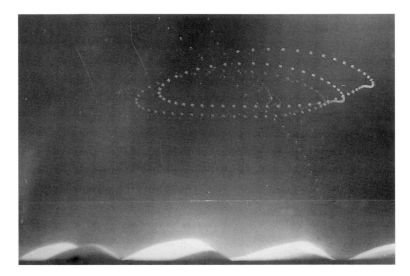

Fig. 9.10 Stroboscopic image to illustrate net forward (left to right) motion of water (from the open, translating ellipses of anticlockwise motion) in the mid-depth of shallow-water waves flowing over asymmetrical wave-formed ripples. Water depth is 1.9 wavelengths, $T = 2.9$ s. (Photo courtesy of K. Tietze.)

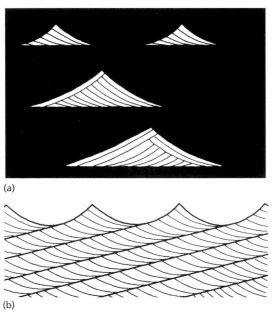

(a)

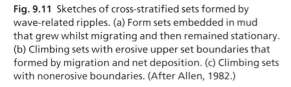

(b)

(c)

Fig. 9.11 Sketches of cross-stratified sets formed by wave-related ripples. (a) Form sets embedded in mud that grew whilst migrating and then remained stationary. (b) Climbing sets with erosive upper set boundaries that formed by migration and net deposition. (c) Climbing sets with nonerosive boundaries. (After Allen, 1982.)

9.4 Combined flows, wave–current ripples and hummocky cross-stratification

As oceanic surface waves propagate on to the shelf, notwithstanding any net shoreward motion of the

kind noted in the previous section, they must coexist there with time-varying currents due to tides. It is not only during local or regional storms that significant waves interact with the shelf bottom and with the tidal stream; a swell of period 10–20 s is usually present on a shelf due to the generation of storm waves in the far-distant field of an adjacent ocean (e.g. Grant & Williams 1984). In the first instance we may look upon the two energy sources as additive; the oscillatory wave motion decaying in amplitude downwards is superimposed upon a boundary layer caused by (in

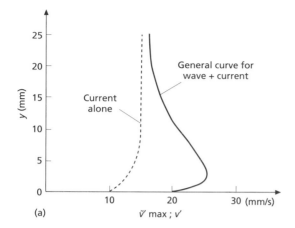

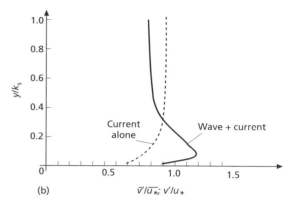

Fig. 9.12 Plot of experimental results for normalized vertical turbulence intensity against dimensionless depth in current alone and in a single run of wave plus current above a rough bed; y is flow depth, k_s is roughness height; for combined flow, $\tilde{v}'/\overline{u_*}$ is ensemble root-mean-square vertical turbulence fluctuation *over* mean shear velocity; for current flow, v'/u_* is root-mean-square vertical velocity *over* shear velocity. Note the very marked increase of near-bed turbulence intensity in the combined flow case. (After Kemp & Simons, 1982.)

the simplest case) a one-way current. Thus at any one point the net velocity field will reflect reinforcement and then opposition as the wave cycle progresses. Experimental (Kemp & Simons, 1982; Murray *et al.*, 1991) and field (Green *et al.*, 1990; Vincent & Green, 1990; Osborne & Greenwood, 1993) data suggest that wave and current effects of combined flows are additive for smooth boundary flows, the boundary layer decreases in thickness and there is an increased velocity gradient close to the bed. The latter observation might lead one to suspect that there is a general increase in the near-bed mean shear stress for combined flows compared to the current alone. Somewhat surprisingly this does not appear to be the case for flows close to the threshold for motion of sands, measurements of bed shear stress at ripple crests indicating values identical to those obtained in current flow alone (Kapdasli, 1991). For gravel beds the combined effects of flow and wave are seen to initiate motion well below the threshold values for waves alone (Panagiotopoulos *et al.*, 1994). For rough boundary flows the effects of wave and current are not simply additive and some sort of interaction is postulated since very high peak vertical turbulent velocity values and increased turbulent stresses occur (Kemp & Simons, 1982; Fig. 9.12). It seems that periodic violent vortices are generated during combined flows close to the rough bed (up to three times the values in pure flows alone), and that the rapid fluctuation of the turbulent stresses arising from these during the wave cycle lead to a marked

increase of near-bed suspended sediment advected up into the flow. It is probably this effect that enables the generation of sediment-stratified flows and of powerful shelf gradient currents during storms (Chapter 25).

Preserved ripple forms generated under gentle combined flows above the threshold for movement may be distinguished by the ripple index (Reineck & Wunderlich, 1968), wave–current ripples rarely showing values greater than about 15 whereas current ripples may show values up to 40, and by the fact that current ripples rarely show crest bifurcation. In environments subject to both pure oscillatory flow and periodic net landward flow, a complex alternation of internal cross-laminations may result.

The wave-formed and combined flow ripple features discussed above have all been formed under laboratory conditions of wave generation appropriate for relatively short-period, two-dimensional surface water waves (< 3–4 s). For very fine to fine sands ripple wavelengths are of the order of 5–10 cm for water wave periods of 2 s or so. Yet we are all aware that during storms great waves arise whose periods may greatly exceed 10 s and whose surface shapes are very far removed from two-dimensional. What bedforms might arise on sand bottoms under such conditions? The answer to this question came initially, not from sedimentological fluid dynamic experiments, but from good 'old-fashioned' geological field observations. J.C. Harms and coauthors (1975) observed that in sandstones of undoubted marine origins there often

Fig. 9.13 Hummocky cross-stratification at outcrop. Middle Jurassic Scarborough Formation, Yorkshire.

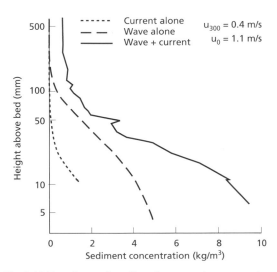

Fig. 9.14 Experimental profiles of wave-cycle-averaged suspended sediment concentration for current alone; wave alone; and wave plus current. It can be seen that the former two are not simply additive to get the latter. (After Murray *et al.*, 1991.)

occurred curious patterns of strongly three-dimensional cross-stratification that they named hummocky cross-stratification (Fig. 9.13). On bedding planes in the geological record hummocky bedforms preserved as form sets appear as convex-up irregular domal features separated by broad three-dimensional troughs. Internally, sections reveal steep to very gentle (most commonly in the range 5–15°) laminations concordant with the exterior forms of dome and trough, but frequently showing erosional low-angle surfaces, onlap and pinchouts.

It was suspected from the occurrence of these enigmatic structures in graded beds that begin in undoubted upper plane bed laminations and end (via the hummocks) in smaller oscillation ripple cross-laminae that the hummocks probably arise during the development and waning of near-bed sediment transport during the passage of storm waves. Experiments (Southard *et al.*, 1990) reveal that, as wave period is increased, ripples become progressively longer in wavelength and also more strongly three-dimensional. At wave periods of 10 s or more the ripple wavelength increases to greater than 1 m. Under such conditions, particularly in strongly combined flows, there is a huge amount of sediment entrained into the flows (Fig. 9.14) (Murray *et al.*, 1991). The sediment-charged

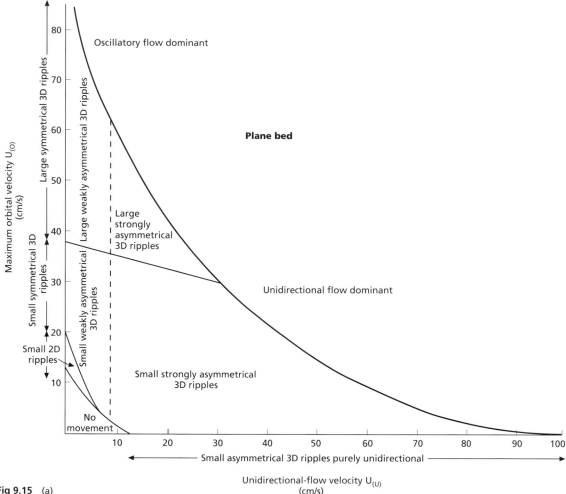

Fig 9.15 (a)

water and strongly 3D forms probably forming the nuclei for the development of hummocky stratification as flow conditions waned, although we still have little knowledge of the essential fluid mechanical details of the process. Myrow and Southard (1991) provide useful conceptual diagrams (Fig. 9.15) that help to shed light on the sequences of bedforms and structures produced during various scenarios of waning unidirectional, oscillatory and combined flows. We may finally note that the high concentrations of suspended near-bed sediment and the occurrence of storm-surge-related gradient currents (see Chapter 25) together cause development of shelf turbidity currents during storms.

9.5 Tidal flows

The periodic rise and fall of sea level visible around coastlines has long fascinated both scientist and mariner alike. For example, it is recorded that the conquering Romans were initially appalled at the large tidal ranges that their navies would have to deal with in the Channel between northern Gaul and Britain and the army refused to budge. Data from tidal gauges reveal a variety of types of tides and important modifying factors arise from the irregular distribution of land and sea, from the irregular shape and variable depth of the sea, from the Coriolis force produced by the Earth's rotation, from the inertia of

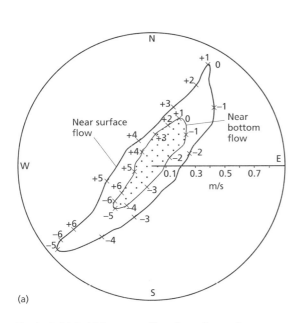

(a)

(b)

Fig. 9.16 (a) A tidal current ellipse from the southern North Sea. Outer ellipse for near-surface flow, inner ellipse for near-bottom flow. The ellipse is constructed by drawing a velocity vector for a given constant height above the bed at various stages (0, +1, +2, etc.) in the tidal cycle. If the complete cycle lasts 12 h then each vector is drawn at 360/12 intervals. It can be seen that tidal currents vary in both time and space, i.e. they are both unsteady and nonuniform. Residual currents may arise because of the asymmetry of the ellipses. (b) Arrays of tidal current ellipses (1 m above bed) for the southern North Sea. (After McCave, 1971.)

Further reading

P.A. Allen (1997) is good on waves and sediment transport. Sleath (1984) is a fundamental (and difficult) reference on sea bed mechanics. Haworth (1982) is a good introduction to tidal currents.

10   Bedforms and cohesive sediment transport and erosion

Throw a man into the Missouri and he will not often drown. It is more likely that he will break his leg.

George Fitch, *American Magazine*, 1907

10.1 The 'special' case of clays and cohesive beds

In our discussions of sediment transport and bedforms thus far we have assumed that beds of stationary sediment grains behave as granular solids, each grain unaffected by the proximity of neighbouring grains. Furthermore we have also assumed that during transport the grains are unaffected in themselves both by turbulent stresses and by the chemical composition of the ambient fluid. None of these assumptions is true for clay minerals, clay mineral flocs and cohesive clay beds. The behaviour of clays in transport follows rules established for colloidal suspensions. We have already discussed the effects of clays on drag reduction in natural flows.

1 Clay mineral grains are small enough for short-range van der Waals–London atomic attractions to bring individual clay platelets together, so that a deposited bed of clay has considerable cohesive strength that must be overcome before erosion of the bed can occur by turbulent stresses.

2 Many clay minerals (particularly kaolinite, illite, mixed-layer clays) carry a net negative electrical charge on their basal faces caused by isomorphous substitution in their lattices of one kind of ion for another. In montmorillonite Mg^{2+} may substitute for Al^{3+}, giving a net unit charge deficiency per substitution. Similarly, Al^{3+} may substitute for Si^{4+} in kaolinite. The approach of two clays in face-to-face contact during turbulent transport *in fresh water* will therefore lead to particle repulsion, even with weakly held cationic species in attendance. On the other hand, attractive forces between clay particles depend upon the existence of a small net positive charge at the broken edge of a clay platelet where metal–oxygen bonds are broken. When particles are brought very close together during turbulent transport, this edge charge participates in an edge-to-face linkage of electrostatic type forming a loosely packed aggregate in a 'house-of-cards' arrangement. The probability of this attraction, as opposed to face-to-face repulsion, is relatively low.

3 Seawater, by way of contrast, is a strong electrolyte, so that the negative face charges are neutralized with respect to the ambient fluid through the effects of cations, which cluster tightly adjacent to the faces to form what has been termed an electric double layer. Because of the very high concentrations of ions (like Na^+) in the ambient fluid, there is now little gradient of electrical potential (the zeta potential) and so approaching clay particle faces (the majority of the 'target' area) no longer repel, allowing van der Waals–London atomic forces to establish attraction. This is the process of flocculation, important in estuaries at salinities of only a few parts per thousand, whereby larger aggregates of grains are formed from a myriad of individual platelets. There is an interesting analogy here with blood coagulation and its facilitation by the addition of ions like Al^{3+} in alum. Flocculation becomes important at the estuary head where fresh water is diluted by turbulent mixing with seawater. The effect leads directly to the production of 'turbidity maxima' and sometimes 'fluid mud', topics discussed more fully in Chapter 21.

4 The bringing together of clay platelets to form flocs is exceedingly complicated. At the smallest scale in near-stationary fluids, collisions between settling grains are due to Brownian motion or the differential settling of variously sized particles. In laminar

boundary layers, collisions are due to differential shear arising from velocity gradients. In turbulent boundary layers, flocs approaching a certain critical size corresponding to the local scale of microturbulence are continuously broken down by turbulent stresses in the zone of highest turbulent production, only to flocculate once more higher in the boundary layer where both velocity gradients and Reynolds stresses are less (see, for example, Jeffrey, 1982; van Leussen, 1997).

5 Finally, flocs are intimately associated with organic matter, the 'sticky' varieties of which, particularly in seawater, may encourage binding and flocculation into large (up to several millimetres) but weak aggregates independently of short-range electrostatic effects.

From the remarks above, it can be appreciated that the critical conditions for entrainment of mud from beds exposed to fluid shear stresses is a complex function of clay type, fluid chemistry, state of fluid flow, organic concentrations, etc. For example, following our discussion of flocculation, we would expect the critical erosion rate for muds to be a very sensitive function of electrolyte concentration. This is confirmed by studies in which various concentrations of $NaNO_3$ are added to the deionized porewaters of pure kaolinite muds. The added salt greatly increased the critical erosive stress (Raudkivi & Hutchinson, 1974). Consolidation due to compaction is also of great importance, leading to an increase in cohesiveness and a decrease in erodibility with depth so that surface erosion by flows may be followed by bed stability at depth (Johansen *et al.*, 1997; Fig. 10.1). Analysis of consolidation experiments done over different time intervals by these last-named authors yielded the general empirical equation:

$$E = k(t)^\alpha (\tau_0 - \tau_c)^\beta \qquad (10.1)$$

where E is erosion rate, t is consolidation time (after deposition), and α and β are experimental constants.

The growth of surface algal films and coatings on many marine muds (clastic *and* carbonate; for case histories see, respectively, Frostick & McCave, 1979; Scoffin, 1970) and their binding effects emphasize the additional role of biological stabilization (see review by Paterson, 1997).

In view of all these factors it is therefore impossible to generalize on values of critical erosive stress and transport relations for muds without specifying mud and fluid composition, organic content and previous

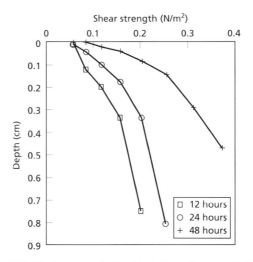

Fig. 10.1 The increase of critical bed shear stress needed to erode deposited Wadden Sea marine muds that have been experimentally compacted and consolidated for different times. (After Johansen *et al.*, 1997.)

depositional history. In particular, experimental data pertaining to freshwater muds are hardly likely to apply to marine situations. In the most general terms it is clear that the erosion of a mud bed by a current will occur when the bed shear stress τ_0 exceeds the local critical shear resistance τ_c. The erosion may take the form of:

- direct particle-by-particle erosion by pick-up when floc–floc bonds or platelet–platelet bonds must be broken;
- mass erosion due to failure above a plane that may or may not be caused by impact of grains or other objects in the flow;
- re-entrainment at the surface of a freshly deposited and water-rich suspension or fluid mud (see Chapter 21).

The dynamics of mud transport by marine tidal and wave flows (Fig. 10.2) is of great practical importance to problems of siltation and effluent contamination. There are many problems involved in the study of flocculated mud aggregates, not the least of which is the difficulty of estimating floc size from seawater samples, the extraction of which may lead to floc disaggregation. A number of *in situ* techniques have been established in recent years (for a good review see Eisma *et al.*, 1997) in attempts to get over such problems. Central to understanding the process

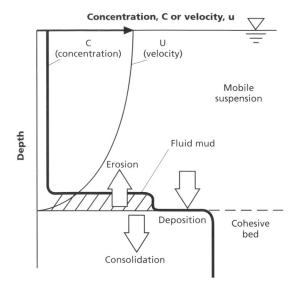

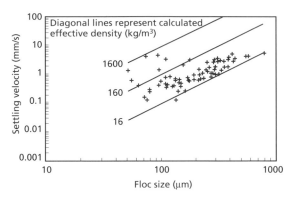

Fig. 10.2 Sketch of processes involved in cohesive bed and mud suspension-related processes. (After Mehta, 1989.)

Fig. 10.4 Settling velocity vs. floc size for flocs measured *in situ* in the Elbe estuary. (After Eisma *et al.*, 1997.) Note the very large range of flocs that have the same settling velocity, and the wide range of calculated effective (immersed) floc densities ranging from values close to pure organic matter (≈30 kg/m³) to values close to solid mineral densities (≈1600 kg/m³).

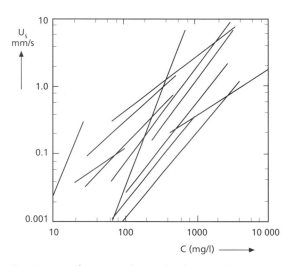

Fig. 10.3 Data from several estuaries showing the variation of settling velocity and particle concentration in mud.

is the concept of collision-induced flocculation noted above. This explains the well-established positive relationship between dilute concentrations of suspended mud (< 10 g/l), the frequency of collisions and the size of flocs (Mehta, 1989). The settling velocity, U_S, of

mud flocs as a function of gross mud concentration, C, for such dilute flows from several estuaries is shown in Fig. 10.3 and may generally be expressed as:

$$U_S = kC^m \qquad (10.2)$$

where k and m are widely variable empirical constants for particular cases. For greater concentrations U_S declines because of the effects of hindered settling noted in Chapter 6 (the Richardson–Zaki formula). Calculations of deposition rates and turbulent diffusion in muddy flows also require knowledge of floc size and density (Fig. 10.4), but these vary widely.

10.2 Flow erosion of cohesive beds

Once the critical erosive stress is exceeded the erosion of a perfectly flat kaolinite mud bed by fresh water proceeds with the development of three sorts of bedform as flow strength is gradually increased.
1 Small-scale longitudinal grooves and ridges have a typical spacing normal to flow of around 0.5–1.0 cm. Well-established examples show that ridges are sharp with broad, rounded intervening furrows. Tiny striae produced by the flow enlargement of air bubbles in the mud indicate a bottom flow structure and scaling consistent with the existence of flow vortices typical of viscous sublayer streaks.

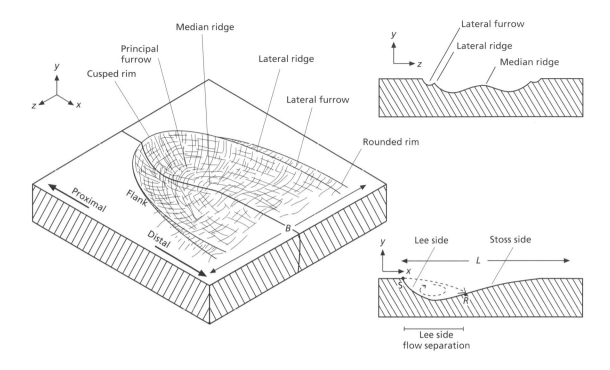

Fig. 10.5 The morphology of an idealized flute mark cut into a cohesive substrate. (After Allen, 1982.)

2 At slightly higher flow velocities the longitudinal grooves change into meandering grooves, indicating a transverse instability affecting the sublayer streaks.

3 The meandering grooves cause deep corkscrew-like erosional marks to form as flow strength is increased. These gradually develop into characteristic spoon-shaped depressions called flute marks (Fig. 10.5). Flutes may also commonly form from bed defects such as hollows or impact marks. They form as a result of flow separation from the lip of the initial hollow. High turbulent stresses at the point of flow reattachment cause deepening and lengthening of the incipient flute. In a mature flute the deepest part usually lies some distance upstream from the reattachment point in the area occupied by a captive recirculating fluid bubble. A large variety of flute shapes occur (Fig. 10.6), depending on initial conditions and the shape of the defects. Flutes range in size from a few centimetres (Fig. 10.7) to several metres

long and wide (Fig. 10.8). The alignment of rows of flutes with intervening flute-free areas shown in Fig. 10.5 may originate from spanwise variations in bed shear stress such as those associated with lobes and clefts at the front of turbidity currents. Little is known about controls on flute size other than the obvious point that large bed impact marks generate large flutes.

It should be noted that each of the above bedforms may be produced by pure fluid stress alone, although most natural examples were probably aided in their formation by the pitting effects of suspended and bed-load particles and by initial bed irregularities such as bioturbation traces.

Laboratory experiments cannot reproduce the spectacular large-scale longitudinal grooves called gutter marks (Whittaker, 1973). These may have a spacing of a metre or more and be up to 20 cm deep (Fig. 10.9). Flutes or tool marks may cover the margins of these features. Intergroove areas may be perfectly flat and show few effects of erosion. Gutter marks are probably caused by the action of the boundary-layer secondary flows but there has been little work on the forces and flows responsible.

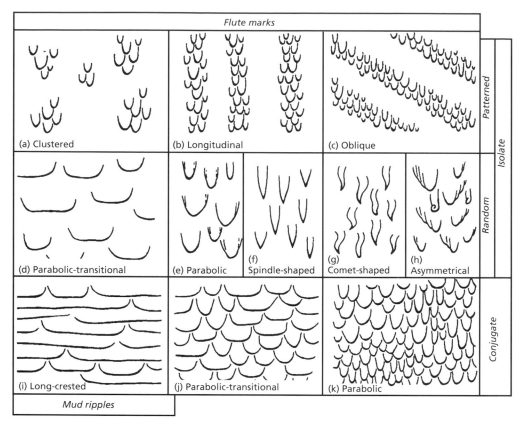

Fig. 10.6 Some of the main types of flute mark assemblages and morphologies found in Nature. (After, Allen, 1982.)

Fig. 10.7 View of base of a Silurian-age turbidite bed showing a nice array of conjugate parabolic flute casts. Palaeoflow was right to left. Horton-in-Ribblesdale, Yorkshire, UK.

Fig. 10.8 The upstream part of a spectacular giant (> 6 m wide and 2 m deep), isolated parabolic flute exhumed on a current-rippled bed top from the base-of-slope storm/turbidite facies of the Carboniferous-age Ross Sandstone, Co. Clare, Ireland.

Fig. 10.9 Sinuous long gutter casts from the base of a catastrophic crevasse-flood sheet sandstone. Middle Jurassic Scalby Formation, Yorkshire, UK.

Fig. 10.10 The base of a thin fine sandstone showing criss-crossing and 'swirling' casts believed by the author to represent tool marks formed by the carriage of plant stems and branches over a partly consolidated mud bed during a river flood incursion into a floodplain. Lower Carboniferous of Dumfriesshire, Scotland, UK. Scale is 10 cm long.

10.3 Erosion by 'tools'

Detritus carried by a flow may form a great variety of impact marks on cohesive mud beds as the 'tools' bounce or are dragged over the substrate. Sometimes these marks may be recognized as caused by a particular tool, as in the case of a saltating orthocone, rolling ammonoid or dragging plant stem (Fig. 10.10). Preserved tool casts may be useful palaeocurrent indicators if the original mark was slightly eroded by flow subsequent to impact, or gave rise to shear drag features such as chevron marks.

Further reading

Burt *et al.* (1997) is an excellent collection of 33 papers on recent studies of clays, clay flocs, fluid muds and much else of relevance to clay sedimentation, settling and erosion. Allen (1982) gives results of theoretical and experimental studies of the origins and characteristics of a variety of erosive bedforms in both muds and rock (including sections on cave scallop solution).

11 Sediment gravity flows and their deposits

Not just the beck only,
Not just the water—
The stones flow also, . . .

Norman Nicholson, 'Beck', *Selected Poems*, Faber

11.1 Introduction and static grain aggregates

Grain aggregates may, with the aid of gravity, transport themselves with no help from any ambient fluid. Let us briefly consider some bulk properties of *stationary* grain aggregates; it is the internal structure, composition and pore fluids of these that determine their reaction to instabilities brought about by gravity or ground accelerations. The accumulation of grains as a deposit inevitably leads to the development of a packing structure that determines many bulk particle properties. The fractional volume concentration (C) of grains within some deposit before cementation is the ratio of grain-occupied space to whole space. The amount of pore space (p) is then:

$$p = 1 - C \tag{11.1}$$

Bagnold (1954a) developed the concept of linear concentration, λ, which is the ratio of particle diameter, d, to the distance between adjacent particles, s, or:

$$\lambda = d/s \tag{11.2}$$

This linear concentration is related to the fractional volume concentration by:

$$C = \frac{C_*}{(1/\lambda + 1)^3} \tag{11.3}$$

where C_* is the maximum possible concentration when $\lambda = \infty$ ($s = 0$) under conditions of rhombohedral packing (see below). C_* is 0.74 for spheres.

The packing of particles in a deposit partly determines porosity, permeability, internal stress levels and stability. Various packing modes for equal-sized sphere assemblages exist. Two simple end-members are cubic and rhombohedral packing (Fig. 11.1), which have porosities of 48% and 26% respectively. These may be regarded as maximum and minimum values for equant spheroids. Many natural granular deposits that have undergone no cementation commonly show porosity intermediate between these values. The distribution of stress in such natural aggregates is of great theoretical and practical interest; it seems that static grain-to-grain stresses are distributed far from equally, with very large internal stresses borne by relatively few grains (for brief reviews see Jaeger *et al.*, 1996; Fineberg, 1996). The implications of such discoveries are yet to be investigated by sedimentologists.

Fig. 11.1 Sketches to show (a) perfect rhombohedral and (b) perfect cubic packing of equal-sized spheres. Any movement of any sphere from close to wider spacing must involve a dilation, Δd.

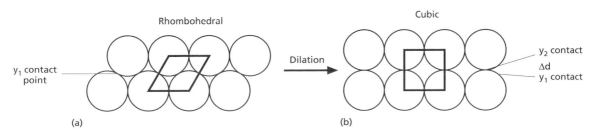

201

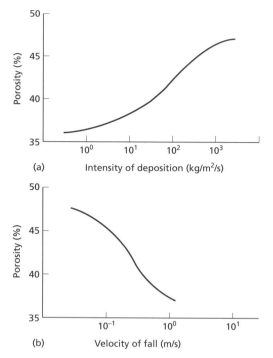

Fig. 11.2 The porosity of sand aggregates as a function of (a) the rate of deposition and (b) the velocity of fall of the grains. (After Gray, 1968.)

A number of factors control the packing geometry and hence the other bulk properties of deposited particles.

1 The porosity of natural sands with similar packing is independent of grain size, but it does vary according to sorting (Beard & Weyl, 1973). Experiments reveal a 25% difference in porosity between well-sorted sands and very poorly sorted sands of the same mean grain size.

2 Grain shape is also important. Consider, for example, the differences produced by the irregular packing of equidimensional spheres, cubes and plates. Very high porosities may be produced during the deposition of shell particles, and 'holes' may be preserved if early cementation prevents compaction. Many freshly deposited clays show very high (up to 90%) initial porosity due to the network structures of clay mineral platelets produced by flocculation, but it is usual for compaction to eliminate this porosity.

3 The rate of deposition affects packing (Gray, 1968; Fig. 11.2) because high deposition rates lead to particle–particle collisions and interference at the upward-moving depositional interface. These effects occur particularly during avalanching down ripple and dune lee slopes and prevent individual particles taking up optimum packing positions, a process that can occur more freely at low deposition rates. Local cubic coordination and the presence of voids thus give rise to increased porosities, a noticeable feature to anyone who has ever trudged home across dunefields.

4 Increasing particle fall velocity tends to decrease porosity (Fig. 11.2) and encourage close packing since particle–bed impacts transfer kinetic energy to underlying layers, causing particle shear jostling and rearrangements that reduce pore space. In fact, vibration-induced resettling of deposited particle aggregates (termed 'granular relaxation' by physicists; see Mehta & Barker, 1991) has many industrial and domestic applications, serving to optimize available storage or carrier space. The process is very important during the resettlement (relaxation) and liquefaction of sand beds by fluidization and other processes.

11.2 Static friction and stability of granular masses

What forces hold aggregates of grains together on a slope? Pour quartz sand on to a smooth quartz mineral surface and also on to a sheet of sandpaper made of the same-sized grains. Tilt the two surfaces carefully at the same rate. Grains on the smooth plate move well before those on the rough plate. Why? In the first case the grains must shear over a surface that is smooth compared to their own curvature, and the frictional resistance to flow comes entirely from the microproperties of the quartz surfaces in contact. In the second case, some grains (a minority, by probability) slide over underlying grains with which they are in cubic contact. The majority nestle between adjacent protruding grains in rhombic contact and must rotate upwards in order to shear. To do this they require more energy, i.e. the slope must be increased by tilting.

In order to quantify these arguments, first consider the effect of surface roughness of the grains alone. The simplest representation would be to consider a mass of grains sliding over the approximately smooth surface of a solid block of the same silicate mineral as the block is slowly tilted (Fig. 11.3). A simple force balance for any given slope, ϕ, gives the normal force,

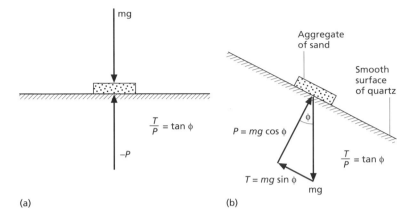

Fig. 11.3 Definition diagrams for (a) normal and shear stresses and (b) the ratio between them, across a plane separating an aggregate of sand grains from a smooth silica surface.

P, as $P = mg \cos \phi$, the tangential or shear force, T, as $T = mg \sin \phi$, and the ratio T/P as:

$$T/P = \tan \phi \qquad (11.4)$$

At some critical value of ϕ the grains slide down the tilted surface. In order to shear the grains over the tilted surface at this critical ϕ, T must exceed a certain limit. The ratio of T/P at this point is thus to be regarded as a friction coefficient, indicating the amount of energy that must be expended to make the grains move. For quartz silt and sand, ϕ varies from about 31° for silt to 22° for pebbles. This is because for any given normal force the load, P_i, *per particle* increases from silt to pebble grades (about 10^4 variation). This discussion of ϕ, sometimes called the *true* coefficient of friction, is *only* of value where we have masses of grains lying upon bedrock slopes. In many sedimentological contexts we must deal with the stability of an already deposited mass of grains that is itself tilted or disturbed from equilibrium, perhaps by tectonic forces. Shear planes may now form *within* the mass of grains, and it is this situation we now analyse.

Consider a volume of dry grains in a container that is tilted at ever increasing angles. At some critical angle, ϕ_{max}, some of the grains will flow off the tilted grain surface as an avalanche or grain flow. The remaining grains are now bounded by a surface resting at a residual angle, ϕ_r, some 5°–15° less than ϕ_{max}. These results are unaffected by the experiment being conducted under water. Attempting to explain the phenomenon (Fig. 11.4) we may first note that the downslope movement of a mass of grains must involve an expansion of the whole mass at failure. This is known as dilatant expansion (another

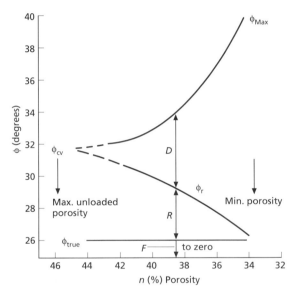

Fig. 11.4 Shear cell experimental results concerning the relation between various measures of the angle of internal static friction, ϕ (ϕ_{max}, ϕ_{cv}, ϕ_{true} and ϕ_r), and bulk mean porosity, $n\%$, for a medium-fine sand. For closely packed sands there is a larger difference between ϕ_{max} and ϕ_r (shown as the difference D) than for loosely packed sands because the shear stress has to do more work in dilating the grain aggregate. The two curves for ϕ_{max} and ϕ_r converge to ϕ_{cv} at high values of porosity. The difference between ϕ_{true} and ϕ_r decreases with increased packing due to less energy spent on grain fabric remoulding (shown as the difference R). (After Rowe, 1962.)

discovery of Reynolds) and it requires energy to be expended (Rowe, 1962; Bagnold, 1966a). Evidently ϕ_{max}, termed the angle of initial or maximum yield, must include the effects of the dilation involved in

moving one grain over another on the zone of potential shear. No such effect is included in ϕ_r, termed the angle of residual shear. The value of ϕ_{max} is strongly dependent upon porosity, which is controlled by the type of particle packing, and it may vary between 40° for tightly packed natural sands and 32° for loosely packed sands (Fig. 11.4). This is because for uniform, tightest possible (rhombohedral) packing, maximum energy has to be done in lifting all grains over their neighbours during dilatant expansion. The stress relation at the point of shear is thus:

$$T = P \tan \phi_{max} \qquad (11.5)$$

Thus in order for a natural grain aggregate to shear, the applied shearing stress must exceed 62–84% of the normal stress due to the static body force. As porosity increases, the difference between ϕ_{max} and ϕ_r steadily decreases, till at maximum (cubic) porosity of about 46% the two values converge to a limiting value. Here there is no need for any dilatant behaviour and we have this limit as ϕ_{cv}, the final or constant-volume friction coefficient appropriate when initial variations in shape and porosity arising naturally through primary deposition are reduced by remoulding and repacking during shear. Grain shape is also an important variable in determining ϕ_{max} and ϕ_r, for angular aggregates of gravel on scree slopes can stand at very much steeper angles (up to 60° for loosely packed deposits) than those of beach, dune and river sands or rounded pebbles.

11.3 Grain flow avalanches: from cross-bedding to megabreccias

Grain flow avalanches are characterized by rapidly moving grain aggregates and by shear at a boundary layer between the moving grains and any stationary underlying and lateral surfaces. If air is the interstitial fluid it plays no, or little, part in the flow dynamics of avalanches and solid-to-solid collisions dominate, at least in the shearing boundary layer. Under water, grain flow avalanches are less 'effective', in that the interstitial water severely dampens the exchange of momentum experienced during grain-to-grain and grain-to-bed collisions. This has little effect on the ubiquitous occurrence of avalanches on the lee slopes of bedforms like subaqueous ripples and dunes, but it does restrict the importance of large-scale subaqueous rock avalanches (but *not* debris flows).

Many material aggregates can avalanche, from a huge range of manufactured products to natural sediment grains and snow. The physics of grain flow and granular materials, previously considered an arcane hobby for sedimentologists, has been much studied and commented on in the past decade (e.g. Barker & Grimson, 1990; Maddox, 1990; Jaeger *et al.*, 1996; Fineberg, 1996; Makse *et al.*, 1997, 1998) because of applications in non-linear dynamics, packing technology (the so-called 'physics of muesli' problem) and in models for low-temperature gases, liquids and solids analogous to simple kinetic theory.

Friction reduction within small subaerial grain flows is probably minimal, but larger subaqueous examples (and powder snow avalanches: see Hopfinger, 1983) may entrain appreciable amounts of fine-grained muddy sediment (or wet snow) that may act to reduce friction and buoyancy contrast, causing the grain flow to become a turbidity or debris flow. Subaerial avalanches may evolve from catastrophic rockfalls and thus tend to occur only on steep subaerial or subaqueous slopes that exceed the angle of slope stability for the particular grains involved. Avalanches occur in a wide range of mountainous terrains and the deposits have a high preservation potential if they should run out from a faulted hinterland into an actively subsiding basin; sometimes it is the faulting that causes avalanching.

Sandy grain flows are usually no thicker than a few centimetres. Coarser grained flows and those containing fine-grained sediment in a subaqueous environment may be much thicker. A common occurrence in modern mountainous terrains and in the deposits of active continental sedimentary basins, particularly rifts (e.g. Burchfiel, 1966; Yarnold, 1993; Friedman, 1998), is coarse sheet-like units termed *megabreccia* derived from catastrophic rockfall avalanches, a group of phenomena termed 'sturzstroms' by German speakers (see Hsu, 1975). These rock avalanches travel at phenomenal speeds (up to 100 m/s) and may ride up over substantial topography during their passage. The majority owe their initiation to the effects of large earthquakes on steep rocky slopes of the basin hinterland. The fact that such deposits are often found large distances from their sources forces us to reconsider the phenomenon of 'runout', defined as the ratio of the vertical fall distance of the avalanche to the subsequent horizontal travel distance. This ratio defines an apparent friction coefficient. For small

Further information on dispersive pressure and grain flow

After slope failure any resulting grain flow will consist of a multitude of grains kept aloft above a basal shear plane. Equilibrium demands that the weight stress of the grains be resisted by an equal and opposite stress arising from the transfer of grain momentum on to the shear plane. Grain flow dynamics are best understood by analogy with basic kinetic theory.

Consider a vessel filled with air and closed by a perfectly sealed but movable weighted piston. At equilibrium the weighted piston is held up by air pressure; the weight force is balanced by a dispersive force due to molecular bombardment by randomly moving air molecules. In one direction there is a constant weight force and in the other there is a net balancing force arising from molecular collisions. Now in grain flows where collisions between grains dominate *throughout the flow* (and there is considerable debate about whether this is generally so), the dispersive normal stress P must be in equilibrium with the weight stress W *at each level*. From experimental data from sheared granular aggregates in a rotating drum Bagnold (1954a) found P to vary as:

$$P = k\sigma\lambda^2 D^2(dU/dY)^2 \cos\alpha \tag{B11.1}$$

where k is a constant, σ is grain density, λ is linear grain concentration, D is grain diameter, dU/dy is the strain rate in the flow, U is the solid velocity and α is an angle of *dynamic* friction. W is more straightforwardly derived from first principles as:

$$W = \bar{C}g(\sigma - \rho)(Y - y)\cos\beta \tag{B11.2}$$

where $\bar{C}$ is the mean grain concentration by volume above point y in the flow, ρ is the fluid density, Y is the flow thickness and β is the local bed slope. Since $P = W$ at equilibrium we may equate (after Lowe, 1976) and solve for dU/dy. Integration with the constraint that $U = 0$ at $x = 0$ gives:

$$U = \frac{2}{3}\left[\frac{\bar{C}g(\sigma - \rho)\cos\beta}{\rho k \cos\alpha}\right]^{0.5}\frac{1}{\lambda D}[Y^{1.5} - (Y - y)^{1.5}] \tag{B11.3}$$

Solving U for various heights in the flow ($\bar{C}$, λ, α assumed constant with height) yields a velocity profile showing a thick shearing zone overlain by a superficial 'plug' of nonshearing flow. This result of Bagnold's was based on experimental measurements of normal and shear stresses caused by neutrally buoyant spheres sheared in a fluid in the annulus of concentric rotating cylinders—the so-called Couette flow. More recent work on the shear of bulk solids indicates that the zone of plug flow (Savage, 1979; Campbell, 1989; Cleary & Campbell, 1993) is considerably thicker than predicted and that Bagnold's experimental artifice may not be an appropriate one with which to model natural avalanches. Nevertheless, the existence of normal stresses (dispersive stresses) during grain flow is confirmed by the simple experiment illustrated in Fig. B11.1.

continued on p. 206

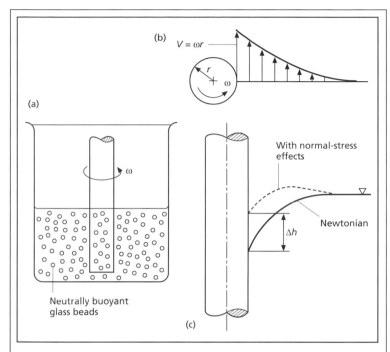

Fig. B11.1 Experiment to indicate presence of normal grain stress (dispersive pressure) effects. (a) Rough-walled rod rotating in a beaker containing neutrally buoyant glass beads suspended in a fluid. (b) Anticipated velocity profile due to the rotating rod (plan view). (c) Increase in free-surface height due to normal stress effects. On the basis of Bagnold's (1954a) experiments it may be estimated that $\Delta h \approx 0.5$ cm, assuming $\lambda = 12$, $\omega = 1000$ rpm and $r = 1$ cm. (All after Savage, 1979.)

Bagnold (1954a) also found that under dynamic shear conditions $T/P = \tan \alpha$. Thus in order for grain flows to develop $\tan \beta > \tan \alpha$. Since $\tan \alpha$ for quartz grains has a value of 0.5–0.6, we must conclude that the initiation of grain flows is possible only on slopes steeper than about 25°–30°. However, once initiated, the flow may continue on very much shallower slopes.

sand avalanches like those on the lee side of a dune or ripple, the runout is very small and the apparent friction very high. Calculated values of the apparent friction for megabreccias are surprisingly low; the larger the initial avalanche volume, the lower the apparent friction and the longer the runout (not uncommonly up to 10 times the fall distance, giving friction coefficients of 0.1). There have been many explanations offered for this long-runout phenomenon, including the following:

1 Flow lubrication caused by ingestion of basal air under pressure (Shreve, 1968; flymo or hovercraft effect)—this fails because megabreccias are common on planets like Mars where, although runout distances are only 50% of those on Earth, atmospheric density is only 10^{-3} that of Earth.

2 Fluidization by ingested air (Kent, 1966) fails for the same reason.

3 Fluid-like behaviour of intraparticular rock dust created by grain collisions (Hsu, 1975)—this fails because any dense *dry* intergranular powder would decrease the kinetic energy and restitution of clast impacts and cause the flow to decelerate rather than to run out.

More information on runout

> Flows of sediment or sediment–water mixtures that are derived from slopes and fall to lower elevations have potential energy mgh, where m is flow mass and h is the average distance of fall. This energy is transformed to flow kinetic energy during flow but is eventually totally lost (as heat) to the effects of frictional resistance as the flow comes to rest. The total energy lost by flow to frictional resistance, f, over its length of flow, l, is thus $mgfl$. Equating energy provided to energy lost gives:
>
> $$mgh = mgfl \qquad (B11.4)$$
>
> and therefore the efficiency of the flow, $1/f$, is given by the ratio of the runout, l, to the vertical fall, h:
>
> $$1/f = l/h \qquad (B11.5)$$
>
> The great attraction of this formula is that the parameters h and l are relatively easily measured.

4 Acoustic fluidization (Melosh, 1979, 1987) due to violent grain impacts from high-fall flows causes high-frequency pressure fluctuations, thus locally exceeding normal stresses and reducing friction.

There is currently some uncertainty as to the exact internal stress–strain regime inside rock avalanches. Earlier theoretical, computational, field and experimental studies (see Savage, 1979; Campbell, 1989; Cleary & Campbell, 1993) established that avalanche frictional effects are confined to a very narrow basal boundary layer and that the majority of the flow travels as a jostling plug. More recent computer simulations (Campbell *et al.*, 1995) indicate that internal shear is more widespread. Plug flow explains the common observation that, although grain-to-grain collisions do occur in the main body, the side-by-side location of broken fragments (jigsaw effect) prohibits much vertical or horizontal mixing, though observations in some ancient examples (e.g. Yarnold, 1993; Friedman, 1998) do reveal evidence for shear planes well within deposits, supporting the Campbell *et al.* (1995) simulation results. Low friction of avalanches is explained by simple particle dynamics, rather than by any more complicated mechanisms of friction reduction. Violent particle collisions with the bed in the thin basal boundary layer can support the entire nonshearing grain mass, which continually provides the kinetic energy to the boundary layer where it is dissipated by largely inelastic collisions (the coefficient of restitution of rock is small). Once the kinetic energy of the main mass has been exhausted then the

flow comes to a halt. The largest kinetic energy comes from the longest initial avalanche freefalls, for doubling the fall velocity quadruples the kinetic energy available. Additionally, during runout the flow falls over solid rock or relatively smooth substrates of low friction.

Another feature of thinner grain flow deposits such as avalanche sets on dune slopes (*not* megabreccias with thick plug-like zones) is the frequent occurrence of reverse grading, where grain size increases upwards in the deposit. The mechanism is spectacularly shown in Fig. 11.5 and the topic is aptly known as 'Why do Brazil nuts rise to the top?' (see Rosato *et al.* 1987) Two hypotheses for this problem of granular physics have been proposed. The first (Bagnold, 1954a) notes that the dispersive stress is greatest close to the shear plane and that large particles can exert a higher stress (to the square of diameter, see eqn B11.1). Hence the larger particles move upwards through the flow to equalize the stress gradient. The second, termed kinetic filtering, says that small grains simply filter through the voids below larger jostling grains until they rest close to the shear plane and the larger grains must therefore rise (Middleton, 1970; Jullien *et al.*, 1992). A simple test for the rival hypotheses is to shear grains of equal size but contrasting density, since the dispersive stress also depends upon grain density. It is observed that the densest grains do indeed rise to the flow surface. Further experiments with naturally varying grain density and size (heavy mineral layers) reveal a consistent pattern of grain segregation such that at

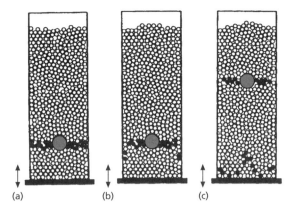

Fig. 11.5 The advective re-sorting of an initial grain configuration due to vibrational resettling and grain-to-grain interactions. Diagrams of the cross-section of the cylinder. (a) The initial configuration before the onset of tapping. A large bead rests in a layer of coloured small beads. This layer is placed in a background of small beads that are identical except for their colour to small beads in the layer. (b) After just one or two taps, the coloured beads against the wall start to move downward. (c) After more taps, both the large bead and the small coloured beads at the centre move up in the cylinder. The beads near the walls that have reached the bottom of the cylinder move inwards and start to rise. (After Knight et al., 1993.)

any level in a deposit the denser grains are smaller than the lighter grains (Sallenger, 1979). This is in accordance with the predictions that at any level in a flow P will be equal. Although the dispersive stress hypothesis is confirmed by such observations, it is still obvious that kinetic filtering is the chief mechanism for sorting and grain migration in multi-sized grain flows, the commonest situation in Nature.

Kinetic filtering plays the dominant role in the size-separation phenomena reported by Knight et al. (1993; Fig. 11.5; see also review by Jaeger et al., 1996), whereby a vibrated mass in a container of equal-sized grains containing one larger grain showed an advective behaviour, with small grains continuously migrating down the walls of the container, and the larger grain and other smaller grains moving up the centre in an advection roll. Such patterns may explain the frequently observed 'patterned ground' produced during earthquake-induced ground vibrations (see Section 12.2). Even more intriguing patterns arise at the free surface of vibrating grain aggregates, the newly discovered 'oscillons' creating much interest amongst physicists in 1996 (see Umbanhowar et al., 1996).

Marked downslope variations in sorting and grain size also develop spontaneously, exemplified by the avalanche deposits of Plates 5 and 6 (Makse et al., 1997; Pouliquen et al., 1997). It is common for larger grains to be carried further than smaller grains because they have the largest kinetic energy. This leads to lateral (downslope) segregation of grain size. More interestingly, when the larger grains have higher ϕ, the mixture spontaneously stratifies as the smaller grains halt first and the larger grains form an upslope-ascending grain layer above them. The reader is referred to the papers by Makse et al. (1997, 1998) for details. Such studies of the granular physics of avalanching are expected to have a major impact upon the interpretation of natural avalanches such as those responsible for the advance of dune bedforms (Fig. 11.6).

11.4 Debris flows

Debris flows are slurry-like flows of diverse grain size, concentration, velocity and internal dynamics. They are equally abundant, and their dynamics are essentially the same, in both subaqueous and submarine settings. In contrast to grain flow avalanches, where solid-to-solid stresses dominate, debris flow involves transmission of both solid and fluid stresses. They are an extreme form of hyperconcentrated flow in which silt- to boulder-sized grains are set in a matrix of clay-grade fines and water. Although the matrix has 'strength', and such non-Newtonian behaviour can support certain grains and clasts, it is also (and some would say mostly) clast buoyancy relative to this matrix and intergranular collisions that support the grains (Pierson, 1981).

Debris flows are probably the single most destructive and dangerous of all the various types of gravity flows. Not only are debris flows common in all climatic regimes but they may occur on a variety of very gentle subaerial and subaqueous slopes, being often funnelled down pre-existing fluvial valleys and channels. This destructive ability comes about because of the combination of extreme mobility (velocities > 10 m/s are not uncommon) and the large impulsive load that may be brought to bear upon any obstacle in their path (their bulk density is in the range 1800–2300 kg/m³). This lethal combination arises because the presence of densely packed coarse solids in a small amount of watery clay matrix (often less than 10% by volume) is conducive to the build-up of excess pore

Fig. 11.6 A typical scene in the lee side of a desert dune, showing the wrinkled surface of grain flow lobes that have descended to the toe of the steep (≈30°) slope after avalanching. Mali Sahara. (Photo courtesy of I. Davidson.)

pressures, which reduce shear strength so that, despite the angularity of the clasts, overall frictional retardation is low to very low. In addition, watery clay mixtures serve to lubricate grain contact surfaces. We may quantify the 'runout' effect in the same way as we did for dry grain flow avalanches: volume for volume, debris flows usually have the greater runout. Data assembled by Iverson (1997) indicate that the ratio l/h is roughly fixed for small flows at about 2–3, but that it increases for larger flows above values of 20 as roughly the logarithm of flow volume.

Subaerial debris flows are common in most climatic regimes and are usually initiated after heavy rainfall from steep debris-laden catchment slopes or vegetated slopes with well-developed soil layers. Natural forest fires act as powerful triggers for debris flows into catchments and, following the great Yellowstone Park fires of 1988, are increasingly seen as an integral part of landscape and sedimentary evolution (see Meyer *et al.*, 1992). Debris flows are of particular importance in volcanic areas when the torrential rains or snowfield melting that frequently follow eruptions lead to widespread and catastrophic flows (termed *lahars*, an Indonesian word) sourced from unconsolidated ash on the volcanic slopes. These are usually more destructive than the direct products of the volcanic eruptions themselves. Historic examples are well described from Mount St Helens (Pierson & Scott, 1985), Nevado del Ruiz (Herd, 1986) and Volcan Hudson (Best, 1992b). Coastal cliffs and land escarpments cut into mudrocks are also the sites of debris flows. Many avalanches transform themselves, and

Fig. 11.7 The Nevados Huascarán avalanche and debris flow, Peruvian Andes. A large earthquake triggered the collapse of part of the 6654 m high mountain, causing $(50–100) \times 10^6$ m³ of rock to drop 4 km and flow out 16 km to the river Santa, where it was transformed into a gigantic debris flow. (From Plafker & Ericksen, 1978.)

the rivers into which they flow, into debris flows as they run out, perhaps the most spectacular historical example being that of the $(50–100) \times 10^6$ m³ Nevados Huascarán earthquake-triggered avalanche in Peru (Plafker & Ericksen, 1978; Fig. 11.7). Less is known about the initiation of subaqueous flows but submarine slope failures and slides seem to provide most of them. Such failures reflect great sensitivity to changes in sediment pore pressure due to loading or storm wave action and sometimes develop from slumps after earthquake shocks. Downslope sediment transfer on many delta fronts is dominated by debris flows derived from side failure. The wasting of passive margin and accretionary prism submarine slopes may be predominantly achieved by failure (creating tsunamis at the same time) and subsequent debris flow transport into the ocean basins (e.g. von Heune *et al.*, 1989).

More information on the physics of debris flows

Debris flows are non-Newtonian in that they possess a yield strength that increases as the amount of fine-grade material is added to pure water. Shear strength depends upon the matrix property of cohesion and upon the granular friction caused by particle interlocking, in addition to resistance to shear caused by matrix 'viscosity' (Johnson, 1970). We can write the following three-part expression for the internal shear stress of a moving debris flow:

$$T = c + F_N \tan \theta + \mu \, du/dy \qquad (B11.6)$$

where c is the cohesion, F_N is the normal weight stress, θ is the angle of dynamic internal friction (~θ_r discussed above), μ is the viscosity and du/dy is the velocity gradient. Inspection reveals that, as the first two terms on the RHS tend to zero, then the equation reduces to that for shear stress in any Newtonian fluid. For the point at which flow begins the condition:

$$T \geq c + F_N \tan \theta \qquad (B11.7)$$

must be satisfied. We may simplify eqn (B11.6) to:

$$T = k + \mu_B \, du/dy \qquad (B11.8)$$

where k is the yield stress and μ_B is the Bingham viscosity (the condition for flow the occur is now $T \geq k$). Equation (B11.8) is known as the Bingham plastic model for debris flows, first proposed by Johnson (1970). It may perhaps be best understood with reference to a simple slab model for the initiation of flow (Nguyen & Boger, 1992). Generally the mean basal shear stress of *any* fluid or solid flow (see Chapter 5) is given by:

$$\tau = \rho g y \sin \beta \qquad (B11.9)$$

We can imagine that for a debris flow the angle β must exceed some critical limit, β_{crit}, before flow begins. Thus the yield stress may be simply determined experimentally by placing a debris flow mixture on a flat surface and slowly increasing the angle β until flow begins. At this point:

$$k = \rho g y \sin \beta_{crit} \qquad (B11.10)$$

The velocity profile across a flowing Bingham substance shows a marked plug profile ($du/dy = 0$) bordered by narrow zones of very high shear strain. Debris flows may be either laminar or turbulent in nature and it should not be imagined that they are sluggish beasts; velocities of many metres per second are commonly recorded. At high Reynolds numbers plug flow may cease; field and experimental observations indicate cohesive strength reduction, frequent grain collisions and interactions at the flow surface (Takahashi, 1978; Wan & Wang, 1994; Iverson, 1997). Many details of the Bingham plastic model for debris flows remain unclear, in particular the occurrence of any shear-dependent viscosity at the strain rates (5–50 s^{-1}) appropriate for most debris flows. There is a useful discussion of this and other points in Nguyen and Boger (1992), Coussot and Proust (1996) and Whipple (1997).

(a)

(b)

Fig. 11.8 Field photos of debris flows, Mt Thomas, New Zealand. (a) Viscous, laminar flow of mean velocity ≈0.2 m/s, $Re \approx 30$. Channel at constriction is about 2 m wide. (b) Higher-velocity viscous debris flow with renewed turbulence. Mean velocity ≈ 5 m/s, supercritical flow and $Re \approx 3 \times 10^3$. Cobbles and boulders in foreground have segregated out to form levees. (From Pierson, 1981.)

The property of matrix strength gives rise to the leveed margins and overhanging snouts seen in debris flow deposits. A mysterious phenomenon is the occurrence of coarse-grained material in the levees and at the snout, seemingly pushed out from the active flow margins (Pierson, 1981; Fig. 11.8). This has been ascribed to dynamic grain interactions (see Section 11.3, the Bagnold effect), but these should instead encourage the grains to move to the flow centre away from the active boundary layer. It is possible that advective motions of the kinds described from grain flows are responsible for these peripheral concentrations. Debris flow deposits are characteristically very poorly sorted (Fig. 11.9) and have few signs of internal structures because of the preponderance of plug flow, which allows no shear fabric or sorting to develop. Traces of shear fabric may occur in basal and marginal zones. Debris flow surging, now recognized as an integral part of flow dynamics, may sometimes produce recognizable stratification due to fabric superimposition and overriding but usually leaves no trace (see Major, 1997; Fig. 11.10). In an important

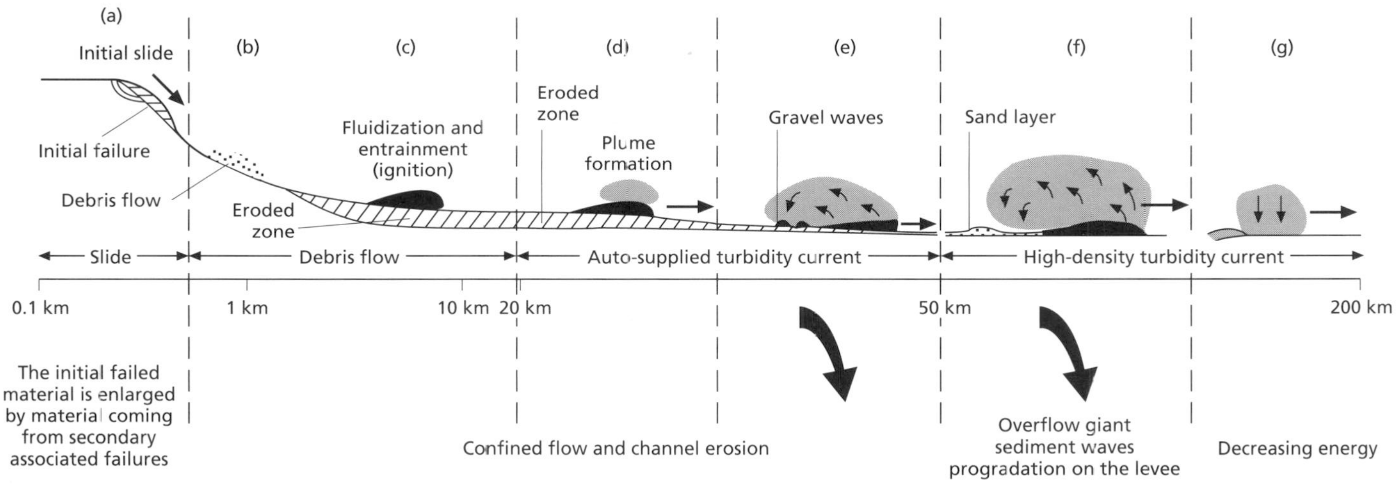

Diagram labels (top to bottom, stages a–g):

(a) Initial slide — Initial failure — Debris flow
(b) Eroded zone
(c) Fluidization and entrainment (ignition)
(d) Eroded zone — Plume formation
(e) Gravel waves
(f) Sand layer
(g)

Slide — Debris flow — Auto-supplied turbidity current — High-density turbidity current

0.1 km — 1 km — 10 km — 20 km — 50 km — 200 km

The initial failed material is enlarged by materia coming from secondary associated failures

Confined flow and channel erosion — Overflow giant sediment waves progradation on the levee

Decreasing energy

Slope gradient	Very high to high	Average between 5° and 1°		Less than 1°			Almost flat
Events	Initialization secondary slides	Disorganization of the slide completed	Fluidization and entrainment (ignition)	Plume formation start of turbulence	Increase of plume thickness		Cable breaks
Average velocity of the current m/s	No current	No current	10–30	7–8	3–5		<3
Thickness of the current (m)	No current	No current	20–30	30–50	50–200	250–100	<70
Channel floor features	Scarps and gullies	Boulder waves			Patchwork of erosional windows		Uniform gray backscatter of the channel floor
Deposit	None	None only large block	Debris flow	None in lows Coarse turbidites in highs	Sand layer at the tail of the current		Silt and fine sand in the channel
Erosion	Limited	Starting	Very active	Active	Decreasing	Limited	None
"Process"	Slide	Debris flow	Dense bottom flow	Turbidity current with plume	Expansion	Massive overflow	Decreasing

Fig. 11.14 The sequence of events that transformed the 1979 Nice airport failure and slide into first a debris flow and then a turbidity current. (After Mulder et al., 1997.)

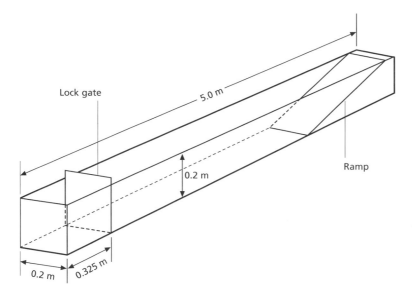

Fig. 11.15 A simple perspex lock exchange tank that is valuable for carrying out experiments into channelized density currents. The ramp is an added extra to study the effects of orthogonal current reflections.

and h is mean height. We might also guess that u should depend directly upon the density difference $\Delta\rho$ between the current and the ambient fluid, or more precisely the action of what we may call *reduced gravity* ($\Delta\rho g$), or g' for short. Experiments (Keulegan, 1957) show that the head of a density current flows with a mean velocity

$$u_h = k\sqrt{\frac{\Delta\rho gh}{\rho}} \qquad (11.6)$$

The constant, k, in eqn (11.6) is insensitive to changes of bed slope but varies inversely with Reynolds number for $Re < 1200$. For $Re > 1200$, $k = 0.7$. It is important to realize that even small slopes cause flow characteristics and morphology to depart radically from the zero-slope conditions discussed in this simple example.

Close examination of the head region (see Fig. 11.16) shows it to be divided into a brain-like array of bulbous lobes and trumpet-shaped clefts (Simpson, 1972). Ambient fluid must clearly pass into the body of the flow under the overhanging lobes and through the clefts. A much greater mixing of denser and ambient fluid by entrainment (about 90% of the total entrainment) also takes place behind the head in the form of periodically shed (and rather beautiful) Kelvin–Helmholtz waves. Internally, the head region shows strongly diverging, predominantly upward mean flow vectors, which help to provide the turbulent fluid

stresses needed for sediment suspension (Middleton, 1966a). The height of the overhang in the frontal lobes is about 10–20% of the total head height for $Re > 100$ (Simpson, 1987).

Continued forward motion of the head at constant velocity requires a transfer of denser fluid (buoyancy flux) from the tail into the head (and thus for the tail to move faster than the head) in order to compensate for boundary friction, fluid mixing and loss of denser fluid in the head region (Fig. 11.16). Little is known about the distribution of velocity and the nature of the boundary layer within density flows. Flow within the head is likely to be highly unsteady and nonuniform because of inflow of tail fluid, entrainment of ambient fluid through clefts and spun-off Kelvin–Helmholtz vortices. It might be expected that mean velocity profiles close to the bed within the body and tail regions should be more like normal logarithmic turbulent profiles (see Simpson & Britter, 1979), but the effect of suspended muds is unknown. Flow of the tail of a current is assumed to follow a Chezy-type law similar to that of rivers (see Middleton, 1966b).

There are two schools of thought concerning the physical nature of 'muddy' turbidity flows. One has them dilute, very thick and sustained (Stow & Bowen, 1980), whilst the other has them, at least in their latter dying stages, concentrated (analogous to fluid muds), very thin and quite brief (McCave & Jones, 1988)! Data from the Holocene of the Madeira abyssal plain

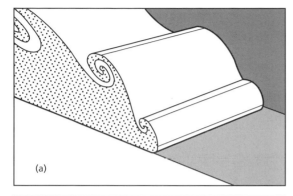

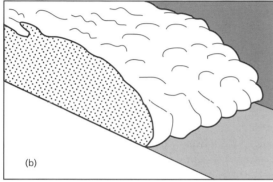

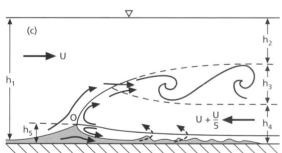

Fig. 11.16 The two kinds of instability at the front of a gravity current head and a section through the head. (a) The developing spanwise Kelvin–Helmholtz waves or billows. (b) The brain-like lobes and clefts that develop at the contact of the overhanging head with a solid boundary. (c) A section through the head with various defined parameters of the head and flow. h_5, height of foremost point of flow; h_4, height of main body of flow; h_3, mixing zone; h_2 ambient fluid; h_1, total depth of ambient fluid; U, velocity. (All after Simpson, 1987.)

cited by the latter authors seem to favour the short, strong and thin idea, although little direct evidence was available to assess flow duration and the extent of flow nonuniformity.

A steady state is brought about in flows that have a near-constant input of dense solution with time. Such flows might occur over a period of time as sediment-laden river-water debouches into a water body and travels along the bottom as a continuous underflow (Prior *et al.*, 1987). By way of contrast, surge-like turbidity flows generated from finite sediment slumps or debris flows on small slopes (< 1°) must decelerate because the supply of denser fluid from behind the head is finite and the buoyancy force driving the flow is insufficient to overcome frictional energy losses (Kersey & Hsu, 1976; Britter & Linden, 1980). The head thus shrinks until it is completely dissipated (but see the box on autosuspension later in this section). On slopes from at least 0.5° to 50° (Middleton, 1966a; Britter & Linden, 1980) head velocity of continuous underflows is independent of slope and varies according to density difference (actually buoyancy flux). Head velocity is approximately 60% of the tail velocity in the slope range 5° to 50°, leading to the head increasing in size as it travels downslope. Entrainment of ambient fluid also causes head growth, increasingly so at higher slopes, and the momentum transferred from the current to this new fluid acts as a retarding force to counteract the buoyancy force due to any increased slope. This 'steady velocity/growing head' behaviour is also a characteristic of starting thermal plumes but has not been investigated from the point of view of turbidity current deposition and erosion. It is expected to play an important role in causing flow overspill onto levees from submarine fan channels.

Rapid dissipation of channelized turbidity flows with consequent deposition must occur as they undergo radial expansion or spreading on entering wide reservoirs or basins. Models for turbidity current deposition are usually 2D and fail to address this important aspect of flow evolution. An interesting transformation takes place for the case of turbid freshwater underflows since deposition progressively reduces the buoyancy force (Sparks *et al.*, 1993). Eventually, as the bulk flow density decreases below that of seawater, the flow may come to a complete halt, with the now positively buoyant fluid rising upwards like an eruptive cloud from a volcano to spread out at the surface as a plume. The process has been termed 'lofting' and may lead to widespread deposition of residual suspended muds over the basin (Stow & Wetzel, 1990).

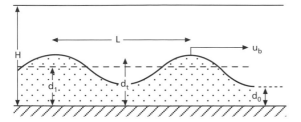

Fig. 11.17 Nomenclature for internal waves moving into a stationary, two-layer fluid: d_1, mean thickness of the bore waveform; d_0, thickness of lower fluid prior to arrival of bore; d_t, maximum height of the bore waveform; H, thickness of the ambient fluid column; L, wavelength of the reflected waveform; u_b, mean velocity of the front of the reverse flow. (After Edwards *et al.*, 1994.)

Fig. 11.18 Nomenclature for a forward flow, the developing bulge at the foot of the ramp obstruction, the three types of bore defined by Rottman and Simpson (1989), and the values for the dimensionless bore height discrimination function found in the experiments of Edwards *et al.* (1994).

Because turbidity flows derive their motive force from the action of gravity, they are easily influenced by submarine slope changes. Flows may partially run up, completely run up and overshoot, or be partially or wholly blocked, diverted or reflected from topographic obstacles (e.g. van Andel & Komar, 1969; Pickering & Hiscott, 1985; Pantin & Leeder, 1987; Muck & Underwood, 1990; Kneller *et al.*, 1991; Edwards *et al.*, 1994; Sinclair, 1994; Alexander & Morris, 1994). The process of run-up and full or partial reflection ('sloshing') is particularly interesting and the effects may be seen by inserting ramps into the kind of lock exchange tanks described previously (Fig. 11.15). Run-up elevations are approximately 1.5 times flow thickness and the process causes upslope deposition on submarine highs (Muck & Underwood, 1990; Lucchi & Camerlenghi, 1993).

Reflection may be accompanied by the transformation of the turbidity flow into a series of translating symmetrical waves, which have the properties of solitary waves or bores (Pantin & Leeder, 1987; Edwards *et al.*, 1994). Briefly, a solitary wave is equivalent to the top half of a harmonic wave placed on top of undisturbed fluid (see Figs 11.17 & 11.18), with all

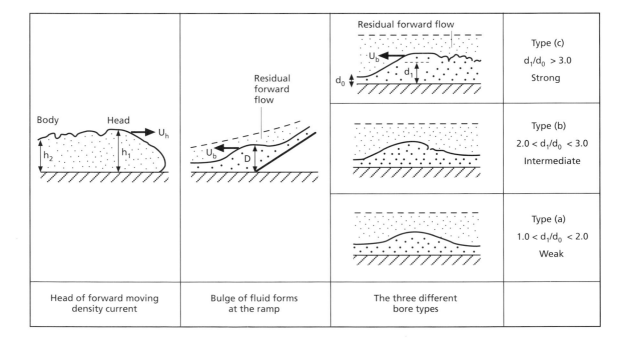

the water in the waveform moving with the wave (bores, unlike surface oscillatory gravity waves, transfer water mass in the direction of their propagation). Somewhat paradoxically we can also speak of trains of solitary waves. They travel back in the up-source direction undercutting the slowly moving nether regions of the still-moving forward current, transporting fluid and sediment mass as they do so. Such bores are termed *internal bores* and they have little vorticity, as witnessed by their smooth forms, in marked contrast to the knobbly 3D nature of the head of a 'normal' turbidity flow. Distinctive deposits result from this process, illustrated in Fig. 11.22. Flow over certain obstacles may also create distinctive lee waves whose properties lead to the formation of regular large-scale bedforms in turbity flow deposits.

Generally, physical analysis of downslope-moving gravity currents (Beghin *et al.*, 1981) reveals that the speed of channelized forward-moving flow decreases as the inverse square root of the flow travel distance (the 'runout'). More recent physical modelling of deposition from 2D surge-like flows that have issued from canyon channels (Dade *et al.*, 1994) assumes a balance between the downslope driving force due to negative buoyancy, its gradual diminution due to particle settling and fluid entrainment, and the effects of friction. This ignores the fundamental point that currents issuing from point sources such as channels must spread radially and hence the 2D analyses are likely to be in considerable error. The 2D analyses may be applicable to flows confined to submarine fan channels but here the common occurrence of prominent large levees implies that frequent overspill of turbidity flows occurs, a process of 'flow-stripping'.

Concerning erosion and deposition under turbidity flows, it is commonly thought that passage of the head is accompanied by erosion, with production of bedforms like flute, groove and bounce, and prod marks characteristically found at the base of many turbidity flow deposits. The extent of this erosion is a little known quantity of some importance in calculating the flux of bed material into a moving flow. Clearly any extra sediment added to the flow will cause the head of the flow to accelerate. Careful biostratigraphic and sediment budget studies of Holocene deposits in the Madeira abyssal plain (Weaver & Thomson, 1993) reveals that 12% or so of turbidite volume is composed of reworked materials. This converts to about

an average of 4.3 cm of erosion over the total areal extent of the deposit.

The study of deposition from experimental surge-like turbidity flows using scaled-down particle density reveals two categories of flow (Middleton, 1966c). Low-concentration ($C < 0.3$) flows show sediment deposition beginning only a short distance behind the head. At any one point, slow deposition from suspension is followed by extended bedload movement and then by rapid deposition from suspension. Finally there is very slow deposition of the finest sediment from the tail. The resulting flow deposits show good sorting, and well-developed upward fining of grain size. By analogy, natural turbidites deposited from relatively dilute flows are expected to show suites of internal sedimentary structures recording diminishing flow strength and changing bedform states with time (the nature of these must be inferred since Middleton's experiments did not produce bedforms as such). In high-concentration ($C > 0.3$) flows, the initial deposition of sediment is followed by mass shearing and formation of a liquefied sediment deposit, which develops instability waves at its upper surface, causing circular shearing motions to extend deep into the bed. As the liquefied bed consolidates, a plane bed surface is formed upon which the finest particles settle from the tail of the flow. The resulting flow deposit shows poor sorting and poor grading.

It should finally be noted that a state of autosuspension may exist in a turbidity flow (Bagnold, 1962; Pantin, 1979). Here, continuous sediment suspension is possible with no internal energy loss.

Turbidity current deposits are called turbidites, with individual turbidite beds ranging in thickness from up to 10 m ('megaturbidite' is an acceptable term for an extraordinarily thick turbidite bed of regional extent) to a few millimetres, and in grain size from coarse gravel to fine silty muds. Their character differs according to mechanism of emplacement (e.g. surge vs. more continuous underflow), entry point and distance from source and position on complex depositional surfaces such as submarine fans or abyssal plains dotted with low-amplitude relief (e.g. Rothwell *et al.*, 1992). But primarily a turbidite sequence depends upon the state of steadiness and/or uniformity of any particular flow. Unfortunately, discussions of turbidite character have been dominated by the concept of 'ideal' sequences of sedimentary

Suspended sediment is held up by turbulence generated at the bed. In normal river flow, gravity causes fluid movement down a slope, which then causes turbulence, which then supports a given mass of suspended sediment. As we have noted above, in a turbidity current the flow is caused by a suspension of sediment, giving a body of fluid excess density, which causes flow down the local gravity slope. Thus a feedback effect arises whence suspension causes motion, which causes turbulence, which causes suspension. This is the state of autosuspension (Bagnold, 1962, 1963). In terms of flow power, autosuspension must arise when the total available power in the current, ω, exceeds the power, ω_t, expended against bottom friction. Thus $\omega > \omega_t$. Put in terms of transporting efficiency of suspended load, the proportion of ω_t available for suspension must exceed the power ω_n needed to support the suspended load. Thus $e_x\omega_t > \omega_n$ where e_x is an efficiency factor. Combining these two criteria for autosuspension and with various substitutions (for a full derivation see Pantin, 1979), we arrive at a condition for autosuspension:

$$\frac{e_x\beta U_s}{V_g} \geq 1 \qquad\qquad (B11.13)$$

where β is bed slope, U_s is the velocity of the suspended load (~ flow velocity) and V_g is the suspended grains' fall velocity. Thus autosuspension is favoured by high efficiency, large slopes and flow velocity, and fine sediment size.

After solving the equations of motion for a two-dimensional turbidity flow, Pantin (1979) and, more generally, Parker *et al.* (1986) show that turbidity currents must either decelerate and deposit their load or accelerate (or 'ignite' in the terminology of Parker *et al.*, 1986) into the autosuspension field where deposition may occur only as the bed slope levels out.

The reader is referred to Pantin's (1979) elegant paper and the mathematically more complete contribution of Parker *et al.* (1986) for a full discussion of this topic.

More about autosuspension in turbidity currents

structures, called Bouma sequences after their originator (Bouma, 1962), but modified in great detail by many subsequent workers (e.g. for fine-grained turbidites see Stow & Shanmugam, 1980; Porebski *et al.*, 1991). In the original 'ideal' sequence of structures in turbidite beds (Fig. 11.19), the following divisions A–E occur (listed from base to top):

A A massive unit whose lack of structure and general grading is thought to represent, by analogy to the Middleton experiments described above, very rapid deposition from a dispersed high-density suspension with much-reduced internal turbulence (see discussion of Kneller & Branney, 1995). Frequent dewatering pipes and dish-and-pillar structures attest to the frequent fluidization present during such sedimentation. Experiments (Arnott & Hand, 1989) also indicate that high deposition rates suppress the occurrence of low-amplitude bedwaves responsible for the laminations seen in division B.

B A planar laminated unit with parting lineations. Usually the laminations are a few millimetres thick and represent deposition from the low-relief bedwaves present on all upper-stage plane beds where deposition rates are relatively low (see Chapter 7).

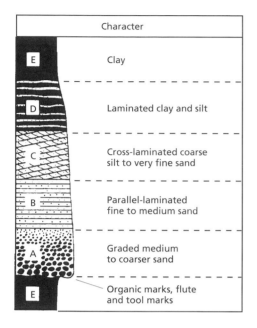

Character
Clay
Laminated clay and silt
Cross-laminated coarse silt to very fine sand
Parallel-laminated fine to medium sand
Graded medium to coarser sand
Organic marks, flute and tool marks

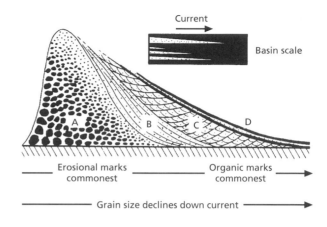

Current

Basin scale

Erosional marks commonest

Organic marks commonest

Grain size declines down current

Fig. 11.19 The traditional view (Allen, 1985) of the Bouma sequence for an 'ideal' turbidite deposited by a waning flow, and the hypothetical variation of such a turbidite sandstone downcurrent.

More widely spaced and inversely graded stratification (Fig. 11.20; given a different code letter by some authors, but this need not concern us) was formerly thought to represent sedimentation of successive 'traction carpets' kept aloft by dispersive granular stresses but now also thought to represent either large-scale variations in turbulent burst/sweep cycles (Hiscott, 1994), current surges or possibly bedwaves analogous to, but larger than, 'normal' bedwaves present on plane beds.

C A small-scale cross-laminated fine to very fine sand unit often of climbing ripple type recording current ripple migration under conditions of net bed aggradation.

D Silts and interlaminated silts and mud sometimes with starved ripples.

E Homogeneous, structureless mud.

The thickness of these various divisions may vary downcurrent, sometimes with a trend for fining and progressive loss of the A–D divisions until only mud is deposited. This is the classic proximal–distal turbidite model originally proposed by R.G. Walker (1965). Studies of Pleistocene and modern turbidites reveal that mud turbidites are considerably commoner, thicker and more widespread than was previously

Fig. 11.20 Examples of the enigmatic inversely graded structure in fine to coarse sandstones from turbidites of Ordovician age, Cloridorme Formation, Quebec, Canada. (After Hiscott, 1994. Photo courtesy of R. Hiscott.)

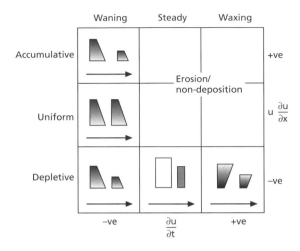

Waning   Steady   Waxing

Accumulative

Erosion/
non-deposition

Uniform

Depletive

$u \dfrac{\partial u}{\partial x}$

+ve

−ve

−ve $\dfrac{\partial u}{\partial t}$ +ve

Fig. 11.21 Acceleration matrix for turbidite beds, with illustrative bed sequences for each field showing downstream and vertical changes in relative grain size of the deposits of each field. (After Kneller, 1996.)

thought. In the South Shetland Trench (Porebski *et al.*, 1991), and the Madeira (Jones *et al.*, 1992) and Balearic abyssal plains (Rothwell *et al.*, 1998), the turbidity currents were probably ponded and reflected within the basins. The great mobility of mud-grade currents must reflect both drag reduction (Best & Leeder, 1993) and the hindered settling of mud-grade particles through turbulent boundary layers.

Until fairly recently the Bouma sequence has habitually been thought to record a decay in flow strength and deposition rate with time at a point as different bedforms equilibrated with a decelerating current. But this is a narrow interpretation and the whole Bouma concept has proved to be a millstone round the neck of geologists whose pattern-seeking tendencies have led to false assumptions about both the nature of deposition and the sequence of events reconstructed for particular turbidity currents. Lest this sound exaggerated, consider the concept of sediment continuity once again (Chapter 6). Recall that deposition, erosion or steady state in a flow can occur

either because the flow is changing power with time at the point of interest or because it has changed power on the way to the point of interest. The first case is that of flow unsteadiness and the second of flow nonuniformity. The two properties are independent and best considered by reference to a simple graphoid (Fig. 11.21).

If a flow is unsteady then it may be either accelerating (waxing) or decelerating (waning). A steady nonuniform flow may be said to accelerate or decelerate only with respect to a moving observer. Fixed observers see no change at a point, but colleagues stationed upstream and downstream would do so (by comparing measurements). The observer in the middle would see tangible evidence for nonuniformity in the form of either sediment deposition (depletive flow) or erosion (accumulative flow). This brings us back to the topic of the Bouma sequence, for it is clear that the ideal succession of structures that has guided field geologists for over 30 years applies to but one very restricted case, that of unsteady (decelerating) flow (Kneller & Branney, 1995; Kneller, 1996). The several types of turbidites summarized in Fig. 11.21 provide very important evidence as to the nature of the turbidity currents themselves, i.e. were the flows steady over a period of time, perhaps hours or days, or were they brief unsteady surges?

Finally, it was briefly mentioned above that flows may suffer reversals or oblique diversions due to the effects of topography. These give rise to very characteristic grain size profiles of repeated coarse pulses with fine drapes and evidence of directional deviance from sedimentary structures, sometimes at 180° but usually less than this (Pickering & Hiscott, 1985; Kneller *et al.*, 1991; Pickering *et al.*, 1992; Haughton, 1994; Edwards *et al.*, 1994; Fig. 11.22).

Further reading

Simpson (1987) is an excellent introduction to the simple experimental fluid dynamics of gravity currents in general. We await a good introductory text on sediment-transporting gravity flows in particular.

Fig. 11.22 (*Opposite*) Bed PF7 from the Cloridorme Formation, Newfoundland, Canada, provides clear evidence for a turbidite deposit produced by initial erosion and deposition by a forward current, followed by the incidence of the flow upon an obstruction, which subsequently generated a series of bores. The sedimentary sequence consists of distinct thin mudstone horizons (II, IV and VI) that separate sandstone packets (I, III, V and VII) in the lower part of the bed. (After Edwards *et al.*, 1994.)

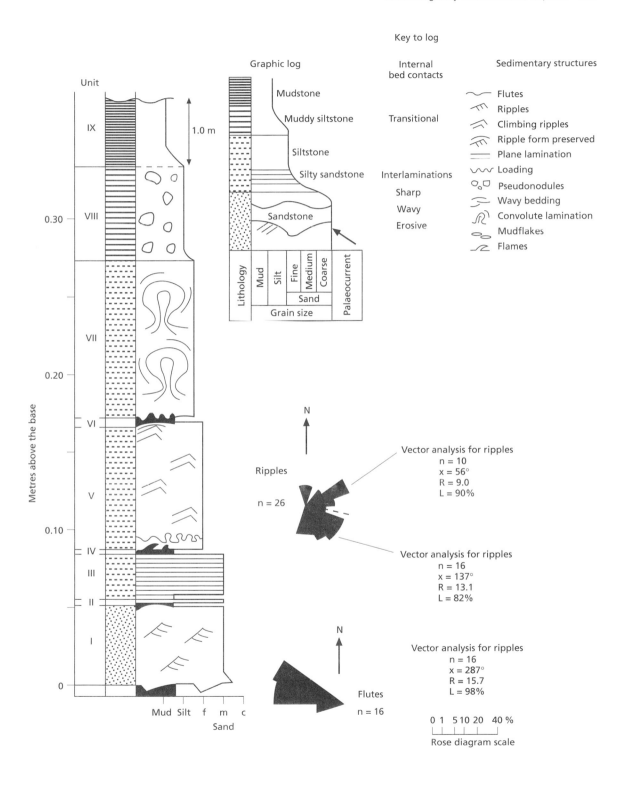

12 Liquefaction, liquefaction structures and other 'soft' sediment deformation structures

The mud, thick as molasses, dripped back into the water with a slow flab flab, and the pole sucked lusciously. It was very beautiful, but it all stank so: yet to his surprise he found he rather enjoyed the rotting smells of the estuary.

Lawrence Durrell, *Mountolive (The Alexandria Quartet), p. 12, Faber and Faber*

12.1 Liquefaction

The stability of a mass of deposited granular sediment like sand or gravel has been seen to depend upon the slope of the mass being less than some critical angle ϕ_{max}. For subaqueous sediment mixtures containing even small amounts ($\gtrsim 0.5\%$) of mud, this critical angle is much reduced to that sufficient to exceed the Bingham yield stress τ_B. Yet for both of these types of sediment there exists a phenomenon known as liquefaction, whereby a stationary submerged mass (or a subaerial one with pore spaces fully water-charged) may be changed *in situ* to a fluid-like state. For example, it is a common observation after large earthquakes that sand and water mixtures are expelled (often violently) from the subsurface in sand 'boils' to form sand volcanoes (Fig. 12.1). Here are the observations made by an observer of the great New Madrid

Fig. 12.1 Line of sand volcanoes (cones are 1–2 m diameter) in a cultivated and irrigated field formed during the 1980 Imperial Valley, California, earthquake.

earthquake of 1811–12 in the lower Mississippi Valley: 'Great amounts of liquid spurted into the air, it rushed out in all quarters . . . ejected to the height from ten to fifteen feet, and in a black shower, mixed with sand . . . The whole surface of the country remained covered with holes, which resembled so many craters of volcanoes. . . .' There is also widespread evidence for the sinking of buildings and other structures into liquefied sands during earthquakes, recalling the Biblical adage that pre-dates modern building regulations by some millenia. Although earthquakes undoubtedly impose an effective trigger mechanism for liquefaction, they are by no means the only one. Thus eruption of sediment and water commonly occurs during river floods from the bases of artificial and natural levees.

What processes lead to production of a liquid-like state from what was previously a stable mass of water-saturated sand? There are two ways in which such a transformation can take place.

1 It may result from the temporary collapse of grain-to-grain contacts in loosely packed sands such as is commonly produced by grain flow deposition. Such a process is known as liquefaction. The energy for granular disaggregation may be provided by the high accelerations experienced during cyclic shock caused by earthquake ground motions. Pressure changes due to the passage of large storm waves over a sandy or muddy (including 'fluid mud') bottom may also cause liquefaction, as can the sudden arrival of a turbidity current or repeated impact of feet on saturated sands on the beach or river bed. 'Vibracoring' devices owe their efficiency to the liquefaction of a narrow zone of sediment around the tip of the core barrel provided by an industrial vibrating tool. Laboratory simulation of earthquake-induced accelerations is possible using shaking tables (Owen, 1996) or a more conventional vibrator. The grains shaken apart are momentarily suspended in their own porewater, causing intergranular friction between grains to be eliminated. Should the liquefied sediment be resting on a slope, however slight, then downslope flow will inevitably result, the flow transforming as it does so into a debris or turbidity flow. On a stationary horizontal bed the situation cannot last and gravity causes the sand grains to settle back once more into grain contact in a new, tighter, packing. This resettlement causes a net upward displacement of pore fluid of volume proportional to the difference in pre- and post-liquefaction porosity.

2 The upward displacement of fluid discussed above may be sufficient in itself to cause overlying grains to remain in suspension by a process known as fluidization or seepage liquefaction. Here adjacent sediment grains are kept apart by an upward force due to the movement of pore fluid along a pressure gradient. This gradient is usually vertical, with p decreasing upwards, but need not necessarily be so arranged. The fluidized suspension may be *en masse* or just restricted to selected pipe-like conduits, often arranged in polygonal forms as described by our acute observer of the New Madrid earthquake. Fluidization in its most simple sense requires the velocity of moving pore fluid, u_f, to exceed the grain fall velocity, U_S. More generally, in order for *en masse* fluidization to occur, the upward-moving fluid must exert a normal stress P_f equal to the immersed weight of any overlying sediment. This force is given by:

$$P_f = \Delta\rho gCh \qquad (12.1)$$

where $\Delta\rho$ is effective grain density, g is gravity, h is layer thickness and C is fractional grain concentration. The conditions for piping are not known since here the upward flow is concentrated into discrete jets, which may separate areas of adjacent unfluidized sediment. Readers may be familiar with industrial and technological uses of fluidization, notably the passage of hot gases through powdered coal beds to optimize burn efficiency. In this case gas fluidization is often accompanied by bubble formation (see Davidson *et al.*, 1977), which is uncommon in liquid fluidization.

Once liquefaction has occurred, the formerly stable aggregate becomes a concentrated grain dispersion of reduced friction, and flow down a gravity slope is therefore possible. The liquefied flow must soon begin to 'settle out' as the grains come into contact once more with their neighbours and displace pore fluid upwards as they assume a tighter packing. Experiments show that a surface of 'settled-out' grains rises up through the dispersion at a rate determined by the fall velocity of the dispersed grains. The flow of a liquefied bed must always be a race against time, the flow gradually 'freezing' upwards with time and becoming immobile. A great variety of collapse and slump structures owe their origins to liquefaction (Nichols *et al.*, 1994; Owen, 1996; Nichols, 1995) and we shall consider these below.

12.2 Sedimentary structures formed by and during liquefaction

No structures result from *en masse* liquefaction since the relative motion between adjacent grains does not occur or is very slight so that depositional fabric remains undisturbed. Only when dewatering fluid is concentrated for some reason do disturbances arise in pre-existing laminations or bedding. The concentrated upward escape of water from a consolidating fluidized bed along conduits causes structureless dewatering pipes or dykes (Fig. 12.2) to form. These may be arranged in crudely polygonal arrays and range from a few millimetres to several metres in height. Any laminations in the liquefied bed serve to outline the shearing effects on the liquefied sand by escaping water. Permeability barriers may be violently punctured during the process (Nichols *et al.*, 1994; Fig. 12.3). Grains suspended by the escaping water may be transported up to the bed surface to form sand volcanoes with diameters of up to 1 m and cone angles of up to 16° (Figs 12.4 & 12.5; Gill & Keunen, 1958). Sand volcanoes are preserved only in quiet subaqueous environments. They are best known in the rock record on the top surfaces of certain turbidite beds and subaqueous slumps. Rare elongate examples provide clear evidence for residual current motion that served to direct the outgoing ejaculations of sand slurry (Fig. 12.5).

Fig. 12.2 Vertical section through a fluidization pipe showing upturned sandstone laminations. Lens cap rests on an orthogonal flat surface.

Fig. 12.8 Brace of listric growth faults in delta-front sediments, Breathitt Group, Kentucky, USA. Large tree is about 10 m high.

on the latter here. Definitions are never our overriding concern in this book, but it is useful to distinguish between a *slide*, still in some sort of stratigraphic order and close to its initial failure plane, and a *slump*, more highly mixed by viscous shear and further travelled.

Concerning the mechanics of slope failure to produce both slides and resulting slumps, the applied shear stress at failure, τ_f, on any slope must exceed the shear strength of the material in question. For shear failure to occur, the Mohr–Coulomb criterion must be equalled or exceeded:

$$\tau_f \geq c' + (\sigma - p) \tan \phi \qquad (12.3)$$

where ϕ is the friction angle, p is the pore pressure and the term $(\sigma - p)$ is the Terzaghi effective normal stress, given by the bulk normal stress, σ, minus the pore pressure. The really important point here is that the Terzaghi effective normal stress may be much lowered beneath the value of σ and the shear strength greatly reduced if pore-pressure regimes are heightened by processes such as fabric collapse, vibrational liquefaction and subsurface pore fluid seepage.

Gently sloping depositional surfaces in cohesive mud deposits are particularly prone to rotational failure; the slip surfaces defining listric faults approximate to arcs (Fig. 12.8) in plan view and have a characteristic downward-curving aspect in vertical section. These may occur on a gigantic scale, as in submarine

(particularly volcanic) escarpments, and may be triggered by major earthquake shocks. Catastrophic continental margin collapses and slides are believed to have occurred over deeply buried planes of weaknesses (*décollements*) located above gas hydrate horizons. Many muddy intertidal point bar and cutbank surfaces show such rotational slides to perfection. In such environments, failure of very gently sloping muddy tidal flat surfaces may even occur if the receding tide can reduce the effective strength by increasing the pore pressure. The resulting flowslides are analogous to the downslope movement of subaqueous fluid muds in estuaries. More important volumetrically are the rotational slides that occur offshore from major deltas. These slides, moving slowly along the fault planes, develop as deltaic clastic wedges prograde out over muddy delta-front deposits in which overpressuring and reduction of effective stress can occur (Crans *et al.*, 1980). The active listric fault is termed a growth fault (Fig. 12.8) because beds on the downthrow sides of such faults are thickened where greater sediment deposition occurs in response to increased subsident 'ponds'. Associated roll-over anticlines are important hydrocarbon traps in areas such as the Tertiary deposits of the Niger and Mississippi deltas (e.g. Weber & Daukoru, 1975).

Gently dipping rotational slides on continental margins and delta fronts give rise to large-volume sediment slumps. Some slumps are clearly triggered by cyclic earthquake shocks acting upon clay sediments of high water content in which porewaters cannot easily escape (undrained condition). The resulting bulbous-nosed slumps with pull-apart structures

Fig. 12.9 View of a prominent slump on a low-angle basal slide plane at about the level of the two students. Slide is overlain by well-bedded turbiditic and storm deposits. Upper Carboniferous, Co. Clare, western Ireland.

and imbrications may slide on their basal fault planes (Fig. 12.9) for large distances on gentle slopes. Such slumps may give rise to true debris flows and, ultimately, turbidity flows if the ambient waters can be mixed into the slump sediment. But many slumps do not mix into debris flows or disintegrate into turbidity currents. Instead, they come to rest at the base of a slope or other negative gradient change where they have high preservation potential.

12.4 Desiccation and synaeresis shrinkage structures

It is a matter of common observation that exposure of wet cohesive sediment to the atmosphere causes the formation of polygonal desiccation cracks as the sediment volume is reduced. These downward-tapering cracks, usually preserved in the rock record as casts on sandstone bases, exist on a variety of scales; the thicker the desiccated layer, the deeper and wider the crack systems. Deep cracks often show the plumose markings seen on rock joint surfaces. Desiccation cracks are rectangular on sloping surfaces such as exposed lake margins (Clemmey, 1976). Desiccation of thin mud drapes or layers may produce mud curls. These have a low preservation potential, tending to become reworked to mudflake intraclasts by subsequent current action. Rarely, wind-blown sand may preserve mud curls *in situ*.

Shrinkage cracks may also form under water following porewater expulsion of synaeresis type, which causes the clayey sediment to suffer a volume decrease. The subaqueous shrinkage cracks thus produced may be single, elongate 'eye-shaped' features or trilete cracks radiating from a central point with no connection to adjacent cracks. The single cracks often show preferred orientations (Donovan & Foster, 1972). Delicate infill of subaqueous shrinkage cracks by sand or silt leads to preservation, but compactional effects often cause the preserved fill to be considerably deformed when seen in sections normal to the former depositional surfaces. The delicate nature of the preservation process means that subaqueous cracks are most common in, and characteristic of, sheltered shallow-water lacustrine environments.

Further reading

Discussions of sediment strength are found in all soil mechanics texts; the reader is especially recommended to peruse Lambe and Whitman (1969) for a clear introductory account.

Part 5
External Controls on Sediment Derivation, Transport and Deposition

13　Climate and sedimentary processes

You don't need a weatherman to know which way the wind blows

Bob Dylan, 'Subterranean Homesick Blues', *Bringing It All Back Home*, CBS Records

13.1 Introduction: climate as a fundamental variable in sedimentology

In this chapter we seek to unravel the complex secrets of Earth's climate and how it has a major effect on sedimentological processes through its control upon surface temperatures, the water cycle and the existence of either greenhouse or icehouse climatic states. What is climate? To the ancient Greeks, κλιμα (klima) was a zone of land with common solar inclination and thus temperature conditions. To me, as a schoolboy, climate meant simply 'average weather'; some long-term (multi-decadal) mean values of variables like temperature and precipitation appropriate to some part of the Earth's surface. More recently we have come to appreciate that the averaging interval needed to obtain climate from weather must be chosen with great care: climate changes and the causes of these are exceedingly subtle and complex, involving interactions between atmosphere, oceans and land. This is because the circulation and dynamics of the oceans are largely driven by atmospheric motions and the transfer of heat energy from the atmosphere; atmosphere and ocean are thus said to be coupled. A major consequence of the examination of Earth's recent and distant history as written in ice, fossils, sediments and sedimentary rocks is that climate has undergone major and sometimes abrupt shifts, leading to such important effects as the shutdown of life under kilometres of ice, major ups and downs of sea level and large-scale perturbations to geochemical cycling, especially that of carbon.

Each climatic variable—temperature, light and precipitation—plays a vital role in earth-surface processes:

1 Temperature and its spatial gradients determine reaction rates in aqueous solutions, photosynthetic primary organic production, and aqueous changes of state by evaporation, condensation and freezing. Temperature tells us information about energy transfer at the Earth's surface, how primary solar radiation is reflected, absorbed and reradiated, and the effects of lithospheric conduction, atmospheric convection and transfer during changes of state. Clearly, temperature levels change in time and space and are remarkably complicated.

2 Light levels control rates of photosynthesis and hence primary production levels. Although direct light levels follow a predictable yearly pattern related to the changing declination of the Sun, they are strongly affected by degree of cloud cover and by atmospheric pollutants like volcanic and desert dust veils.

3 Precipitation falls as rain or snow and may accumulate as ice. Water flux controls rates of aqueous chemical weathering and sediment transport. During the growing season when light is at a maximum, temperature is the major control on terrestrial primary production.

More specific to sedimentology, the climate of a region controls:

- the rate of chemical weathering of silicate minerals according to temperature (via the Arrhenius rate law) and the throughflow of precipitation;
- the amount of surface runoff, which leads to erosion and sediment transport and depends on the residual water left over after uptake by vegetation and subsurface seepage;
- the yearly distribution of runoff, which plays a major role in determining stream channel and floodplain sedimentary processes through its control on the hydrograph, and also on the timing and extent

of sediment, nutrient and freshwater export to and across the shelves into the ocean basins;

- the nature of surface sedimentary environments, chiefly the propensity for chemical precipitation under conditions of high evaporation.

13.2 Solar radiation: ultimate fuel for the climate machine

Our Sun radiates energy into space on account of the exceedingly high temperatures reached at and close to its surface. This solar energy is transmitted throughout the Solar System as electromagnetic waves of a range of wavelengths, from X-rays to radio waves, all travelling at the speed of light. The Sun's maximum energy comes in at a short wavelength of about 0.5 μm in the visible range. Much shorter wavelengths in the ultraviolet range are absorbed by ozone and oxygen in the atmosphere. The magnitude of incoming radiation is represented by the solar constant, defined as the average quantity of solar energy received by normal-incidence rays just outside the atmosphere. It has a value in the range 1360 ± 20 W/m².

The Sun's radiation falls upon a bewildering array of natural surfaces, which partly reflect, absorb and transmit it to varying degrees. As well as a multiplicity of rock surfaces of different surface colour dependent upon mineralogy and surface chemical weathering, there are vast expanses of sandy desert, vegetated forests, grasslands, lakes, oceans and ice-sheets. Each one of these has a different behaviour with respect to incident radiation. Thus solids like ice, rocks and sand are opaque and the short-wavelength solar radiation is either reflected or absorbed. Water, on the other hand, is translucent to solar radiation in its surface waters, although when the angle of incidence is large in the late afternoon or early morning or over a season the amount of reflected radiation increases. It is the radiation that penetrates into the shallow depths of the oceans that is responsible for the energy made available to primary producers like algae. We must be very careful indeed to distinguish between directly reflected incoming short-wave solar radiation and reradiated Earth radiation. The former is reflected as identical short-wave radiation and plays little role in atmospheric heat transfer. We shall see below that the latter is longer-wavelength infrared radiation that can be trapped by the gases of the atmosphere. It is useful to have a measure of the reflectivity of natural surfaces to incoming short-wave solar radiation. This is termed the *albedo* and is the ratio of the reflected to the incident short-wave radiation. Snow- and ice-fields have very high albedos, reflecting up to 80% of incident rays, whilst the equatorial forests have low albedos due to a multiplicity of internal reflections and absorptions from leaf surfaces, water vapour and water droplets. The high albedo of snow is thought to play a very important feedback role in the expansion of snowfields during periods of global climate deterioration.

13.3 Earth's reradiation and the 'greenhouse' concept

Incoming short-wave solar radiation in the visible wavelength range has little direct effect upon Earth's atmosphere, but heats up the surface in proportion to the magnitude of the incoming energy flux, the surface albedo and the thermal properties of the surface materials. It is the reradiated infrared radiation that is responsible for the elevation of atmospheric temperatures above those appropriate to a grey body of zero absolute temperature. The reradiated energy flux of about 350–400 W/m² is of the same order as that received from the Sun at the Earth's surface (for comparison, the amount of heat energy received by any surface from *within* the Earth is only about a maximum of 0.1 W/m²). Some of this energy is lost into space for ever but a significant proportion is absorbed by the gases of the atmosphere and emitted back to Earth as counter-radiation where together with absorbed short-wave radiation it does work on the atmosphere by heating and cooling it. During this process water vapour may condense to water, or vice versa, and the effects of differential heating give rise to density differences, which drive the general atmospheric circulation. The insulating nature of Earth's atmosphere is nowadays referred to as the 'greenhouse' effect and we see that it is the absorption spectra of our atmospheric gases that ultimately drive the atmospheric circulation. Water vapour is the most important of these gases, strongly absorbing at 5.5–8 and > 20 μm wavelengths (see Fig. B13.1). Carbon dioxide is another strong absorber, but this time in the narrow 14–16 μm range. The 10% or so of infrared radiation from the ground surface that escapes directly to space is thus mainly in the 3–5 and 8–13 μm wavelength ranges. We may note in passing that it is

Further information on radiant energy transfers

In order to understand the basic physics behind atmospheric heat transfer, we must realize that radiant energy is emitted from any object whatsoever whose temperature is above absolute zero. The Stefan–Boltzmann law relates the total energy of emitted radiation, E, to the absolute surface temperature, T, in kelvins, of a perfectly radiating body (called a blackbody), viz:

$$E = \sigma T^4 \tag{B13.1}$$

where σ is the Stefan–Boltzmann constant of value 5.67×10^{-12} W/cm^2 k^4. You can see that the total energy emitted varies strongly and positively as the fourth power of temperature. Yet important as the total energy of the solar flux is, it is the wavelength of the radiation that is of equal interest and this also varies with temperature. This dependence is expressed by the Wein displacement law, which states that the wavelength of maximum energy, λ_{max}, is inversely proportional to the absolute temperature, T, times a constant, α, i.e.

$$\lambda_{max} = \alpha / T \tag{B13.2}$$

where α has a value of 0.29 cm K when λ is expressed in centimetres. This determines that the solar energy wavelength maximum is in the short-wavelength band at about 0.5 μm. On receipt of such short-wave solar energy, Earth, whose mean surface temperature is a paltry 300 K compared with the Sun's 5500 K, absorbs and reradiates the energy back into space at a longer mean wavelength of about 10 μm, in the infrared range (Fig. B13.1). It is this property of a planet, that of reradiation, that lies at the bottom of all discussions about global climates, for Earth and for all our other planetary neighbours in the Solar System.

The key concepts regarding radiative energy transfer relevant to climate are that Earth and its atmosphere and hydrosphere:
- receive annual levels of energy in inverse proportion to latitude;
- directly reflect a portion of the incoming energy;
- absorb and reradiate upwards and downwards a further portion, but at different wavelengths.

Regarding the latitude dependence of incoming solar energy, it is a simple fact that oblique incident light can only directly warm a fraction (in proportion to the cosine of the angle of incidence) of the surface area that can be warmed by normally incident light. A further control is that solar radiative energy is progressively dissipated by scattering and absorption en route from the top of the atmosphere downwards. Since light has to travel further to reach all latitudes north and south of any line of normal incidence, it is naturally weaker in proportion to the distance travelled. The fraction of monochromatic energy, δ, transmitted at any moment in the daylight solar cycle along a given path of length, x, through the atmosphere is given by the Lambert–Bouguer absorption law as:

$$\delta = \exp(-\beta x) \tag{B13.3}$$

continued on p. 240

where β is the volume absorption coefficient, whose value at sea level is about 0.1 km⁻¹. For example, after 10 km of travel only 1/e (37%) of any incoming monochromatic energy is transmitted. In addition to this mean absorption of energy by atmospheric gases, radiative energy is also reflected and absorbed by wind-blown and volcanic dust and natural and pollutant aerosol particles in the atmosphere. Since these are often injected by natural highly irregular processes like volcanic eruptions, then it is easy to appreciate that the magnitude of incoming solar radiation may vary strongly with time (by up to 20% or more) at any one latitude. Because of scattering, absorption and reflection of incoming solar radiation, it is usual to distinguish the direct solar radiation received by any surface perpendicular to the Sun at that particular moment from the diffuse radiation received from the remainder of the atmospheric hemisphere surrounding it. Continuous cloud cover reduces direct radiation to zero, but some radiation is still received as the diffuse component.

Further information on radiant energy transfers [*continued*]

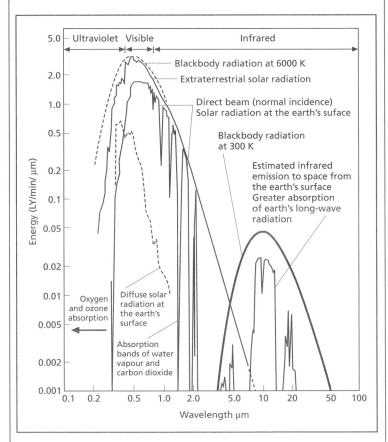

Fig. B13.1 Electromagnetic spectra for incoming short-wave solar radiation and outgoing long-wave (infrared) reradiation; these form the basis of understanding Earth's climate. Curves are labelled to illustrate points discussed in the text. (From Lockwood, 1979, and sources cited therein.)

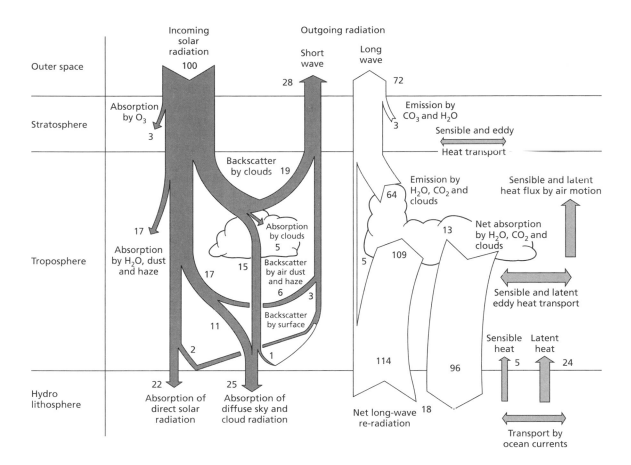

Fig. 13.1 Energy transport into and out of the atmosphere: the vertical fluxes of the Sun's radiative energy at the top of the atmosphere (100 units) and reradiated short- and long-wave radiation. (After Lockwood, 1979, and sources cited therein.)

this escaping infrared (and other wavelength) radiation, which varies with the nature of surface emissivity, that is detected by remote-sensing cameras on aircraft, satellites or high-flying 'spy-planes'.

13.4 Radiation balance, heat transfer and simple climatic models

Let us now consider the net balance of incoming and outgoing radiation over a long time period and how some of the energy is transformed into other forms of energy, like heat, potential and kinetic energy (Fig. 13.1). If we assign 100 units for the magnitude of the incoming short-wave solar radiation (sometimes

termed insolation) flux, then because of the Earth's planetary albedo, 30% of this flux is reflected back into space. Of this total, 6% is reflected from the atmosphere, 20% from clouds behaving as perfect blackbodies and 4% from the Earth's surface. This reflected radiation plays no part in the climate system. The remaining 70% is either absorbed by the Earth's surface as direct and diffuse radiation (51%) or else absorbed by the atmosphere and clouds (19%). The small atmospheric absorption contrasts with the large surface absorption, which is converted into heat energy and thus, by Wein's law, is reradiated back into the atmosphere as long-wave radiation, where most of it is absorbed and then re-emitted (at the same wavelengths), and so on in a complex way. More long-wave radiation in net terms is lost in this process from the atmosphere to space (64%) than is absorbed (15%). Together with the 19% of short-wave radiation absorbed, this means that there is a 30% total *atmospheric* absorption deficit. Should the matter rest

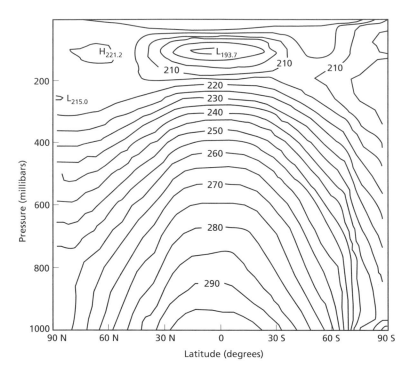

Fig. 13.2 Zonally averaged temperature (K) structure of the whole atmosphere from equator to pole as modelled by a global climate model based on sea-surface temperatures over a 10-year period. Note the deepest area of troposphere occurs at low latitudes. (After Kiehl, 1994.)

there then the Earth's atmosphere and surface would cool drastically to way below 0 °C in no time at all. In fact, the atmospheric deficit is made up by the radiative balance produced by the transfer of heat energy from the surface by conduction and turbulent exchange (sensible heat transfer) and by the formation of clouds and precipitation after evaporation of surface waters (latent heat transfer) (Fig. 13.1). Thus although the atmosphere away from the tropics is in radiation deficit (Fig. 13.2), the overall net radiation from the Earth's surface is positive, so the net radiation to and from the planet is in balance due to the transfer of heat from low to high atmospheric latitudes by convection, precipitation and turbulent transfer.

The simplest climatic case to consider would be if, like our Moon, we had no atmosphere; then at any one time there would be a perfect energy balance between incoming short-wave insolation to the side of the Earth facing the Sun and outgoing long-wave reradiation from the whole Earth. We can write this balance as:

$$0.25(1 - \alpha)S = \varepsilon\sigma T^4 \qquad (13.1)$$

where S is the solar constant, α is the whole Earth's albedo (about 30%), 0.25 is the ratio of sunlit to

whole Earth ($\pi r^2/4\pi r^2$), T is earth-surface temperature, ε is the mean emissivity of the earth in the middle infrared and σ is the Stefan–Boltzmann constant. Solving for T we get about 254 K, or −19 °C. This compares to the actual average mean surface temperature of around 14 °C. The deficit of 31 °C is due to the previously discussed greenhouse effect, whereby Earth's atmospheric gases absorb, reradiate and reabsorb significant portions of the outgoing infrared radiation from the surface and make this energy available for the lateral and vertical transport of heat energy by the atmosphere.

In the absence of an atmosphere or hydrosphere then regions of the Earth would have a simple temperature-variable climate based upon daily, monthly and yearly latitudinal variation of incoming radiation and the albedo, infrared emissivity and radiation loss from the surface. As we have seen, the presence of surface water and atmospheric gases changes this simple Moon-like scenario considerably, since in addition to radiative heat transport the atmosphere may transport heat energy by vertical and lateral movements of atmospheric air masses. It is only in dry desert areas that the climate is dominated by radiative exchanges alone, with high daily and low nightly temperatures:

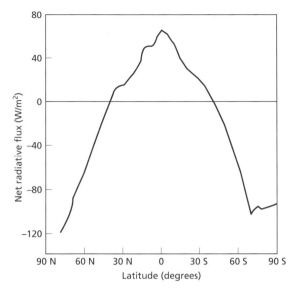

Fig. 13.3 Net annual zonally averaged radiative flux (short-wave absorbed flux minus outgoing long-wave flux) for the top of the atmosphere. (From Kiehl, 1994.)

13.5 Climate and the water cycle

Most of Earth's surface is more or less moist and covered with vegetation, so we must consider the role of the water cycle in the climatic machinery. In such cases another form of heat energy, that of latent heat exchange, is provided when water vapour condenses to rain or snow and is given up when water or ice evaporates. The ratio of sensible to latent heat loss to the atmosphere from a wet surface decreases from about 0.5 to 0 as the temperature increases from 5 to 35 °C. Plants themselves act as agents for evaporation as they transpire water vapour to the atmosphere during daylight hours. The combined effects of physical evaporation from surface waters and plant transpiration is known as *evapotranspiration*.

Earth's oceans have an important role in controlling climate, for the thermal heat capacity of a water mass is very much greater than that of an equivalent mass of air. Thus whilst the total oceanic mass is almost 300 times that of the atmosphere, the combined heat capacity of the oceans is 1200 times as much. Ocean water thus has a considerable thermal inertia, enabling heat energy produced by radiation in low-latitude surface waters to be transferred widely by ocean currents (themselves largely dependent upon atmospheric winds), all the time losing latent heat by evaporation to the overlying winds, which eventually goes towards heating the atmosphere in temperate frontal systems following precipitation (see box below). Climate is thus highly modified by land–sea interactions (Fig. 13.4).

the predominant energy transfer globally is in the form of sensible heat exchanges.

The result of all these heat transfer processes is a mean thermal structure for the atmosphere as shown in Fig. 13.3. The warmest temperatures, about 27 °C, are at lowest latitudes, decreasing towards the poles and vertically up to the top of the tropopause at between 10 and 15 km elevation. The troposphere is thickest at low latitudes, thinning towards the poles. The greatest gradients of T occur at the top of the troposphere.

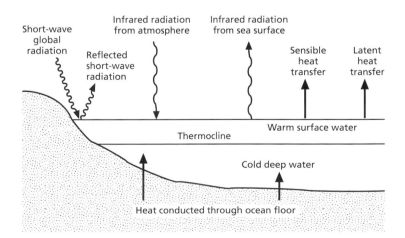

Fig. 13.4 Energy balances appropriate to the oceans. (After Lockwood, 1979.)

As air moves around the atmosphere in response to pressure gradients, Earth's rotation and density differences, energy is continuously being transformed from one form to another. These forms are, overwhelmingly, latent heat energy E_L, sensible heat energy E_S and potential energy E_p (the kinetic energy of air masses is very small and neglected here).

1 Latent heat energy E_L appears when we have moist air, the usual condition for most air masses. It arises from reversible changes of state between liquid and gas, in this case water and water vapour. It is given by the product of the latent heat of evaporation or condensation (L, 2.3×10^5 J/kg) times the mass (m) of water vapour involved. The large volumes of atmosphere involved mean it is very important in atmosphere–ocean coupling. Evaporation of water into a parcel of air requires work to be done breaking hydrogen bonds and hence energy is taken in (from the other energy sources in the atmosphere) and cooling takes place. The opposite holds during condensation of water vapour to rain or cloud droplets and the air heats up.

2 Sensible heat energy E_S is due to the direct impact of radiation exchanges, the atmosphere losing sensible heat by radiation to space and gaining it by radiation from space and reradiation from clouds. It is given by the product of the specific heat capacity at constant pressure, c_p (J/kg K), and temperature, T (K). Water has a very high specific heat capacity (around 4000) compared to that of air (around 1000), for reasons discussed in Chapter 2.

3 Potential energy E_p is that of position, y, above sea level, times gravity.

Our simple equation for the total energy, E, present in any *unit* mass of moist air is thus:

$$E = E_L + E_S + E_p = \text{constant} \tag{B13.4}$$

or

$$E = l + c_p T + yg = \text{constant} \tag{B13.5}$$

Since the total energy, E, is constant, there must be no divergence, i.e. $\nabla \cdot E = 0$.

We are now in a position to understand energy changes in ascending or descending air masses. Such motions dominate heat exchange in the tropics and in all latitudes over elevated relief. First consider a dry, slowly ascending mass in which we have no E_L term. The increase in E_p *must* now be accompanied by a fall in temperature so that E_S can decrease. We can think of this easily in physical terms with a little help from kinetic theory and the gas laws, since as the dry air slowly rises it moves into regimes of lower pressure and hence work must be done in expansion and temperature falls; vice versa for falling air. This change of T with height is known as the *dry adiabatic lapse rate*, about 1 K per 100 m. When water vapour is present in ascending air, then the lapse rate is reduced because as cooling takes place the air becomes saturated, condensation occurs and latent heat is released, becoming sensible heat. This *saturated adiabatic lapse rate* varies with T, being about equal to the dry rate for cold air, but much less for warm air. A typical average value would be around 0.6 K per 100 m.

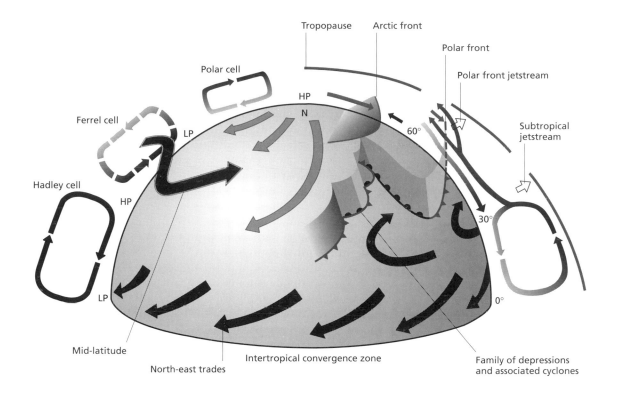

Fig. 13.5 A schematic representation of the general circulation of the atmosphere (northern hemisphere). One family of depressions is shown on the polar front. (From Brimblecombe & Davies, 1982.)

13.6 General atmospheric circulations

General

Vertical movements of the atmosphere are predominantly convective in origin, in that they depend largely upon density differences caused by heating or cooling. Lateral movements are termed *advective* and result from horizontal pressure gradients as strongly modified by the Earth's rotation. These advective motions are seen as surface winds whose stresses act upon the oceans, which respond in the form of ocean currents.

In order to understand the principles of general atmospheric circulation, it is first necessary to see how energy is transported by a unit mass of moist air. We have seen in the box that the total energy, E, is constant, and since E is continuously being reduced by the loss of long-wave radiation to space (the atmospheric radiation deficit discussed previously), the energy is kept in balance by the addition of sensible heat from surface land and ocean and, much more importantly, by the release of latent heat during rainfall. In detail it is found that low latitudes between about 40°N and 40°S have a positive radiation balance, whilst high latitudes have a negative balance. The general circulations of the atmosphere thus involve a poleward transport of heat energy to maintain the observed long-term temperature distributions, which are approximately constant. The simplest possible such arrangement would be a general convective upwelling of warm moist air from the equatorial regions and its transfer towards the poles, cooling by radiative heat loss as it does so, where it eventually sinks, liberating rain, snow and latent heat. Such a simple cellular circulation has the right principles but inevitably the Earth's atmosphere is more complicated (Fig. 13.5), chiefly because of the effects of the Earth's rotation, but also because of pressure effects due to latitudinal nonuniformity in the thickness of the troposphere.

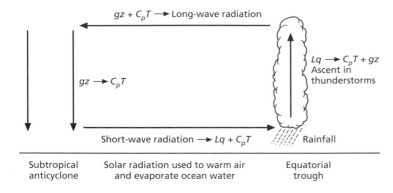

Fig. 13.6 Energy conversions in a simple Hadley circulation cell. (After Lockwood, 1979.)

Low-latitude circulation and climate

In the simplest reduction of low-latitude circulations, between about 30° latitude of the equator, warm moist rising air originates at the equatorial inter-tropical convergence zone, where incoming radiative heating is at its most effective, then travels north and south to sink as cooled air at subtropical latitudes, returning to the intertropical convergence zone at low levels as the generally steady and dependable Trade Winds. The warm low-density equatorial air creates a low-pressure zone at the tropics, which shifts north and south of the equator during the course of the year. The descending cooling high-density subtropical air creates the zonal high-pressure belts of the continental deserts. Such a circulation is termed a *Hadley cell*. In terms of energy transfer (Fig. 13.6) the warm saturated ascending equatorial air, most in thunderstorms, is cooled and latent heat energy is released during rainfall. There is some limit to the height at which the equatorial air can rise, reached when it is cooled and its density approximates to that of the ambient troposphere. The warm dry air now flows polewards, losing further energy as it does so by emission of long-wave radiation gained by its high sensible heat content. We should briefly note here the occurrence of strong upper-troposphere winds, the subtropical Jet Streams, which exist at about the latitudinal limits of the equatorial Hadley cells. These arise because the poleward-moving air masses are accelerated as they move north and south by the Coriolis acceleration necessary for them to conserve angular momentum. Eventually the air in the upper limbs of the Hadley cells must sink and a fairly large proportion of it does so in the subtropical zones with

the transfer of the potential energy of position to sensible heat according to the dry adiabatic lapse rate. Thus the dry cloudless subtropical deserts have their radiation deficits (due to high infrared radiative heat loss from the high-albedo/low-cloud-cover environment) compensated by the lateral transfer of heat from the equator. Over the subtropical oceans the low-humidity air and cloudless conditions enable short-wave solar radiation to warm and evaporate seawater. The Trade Winds then transport the latent and sensible heat in the form of moist, near-saturated winds back to the low-pressure equatorial belt, thus completing the Hadley circulation.

Important seasonal variations occur in the Hadley circulation because of changing thermal gradients, notably a transequatorial migration of the intertropical convergence zone and substantial enhancement of subtropical highs in winter over the cool continents and their diminution and change to deep lows in summer as the land warms up. The subtropical maritime zones of Asia and Africa are therefore noteworthy because of such seasonal reversals (the Coriolis force is low at low latitudes) of the marine-sourced Trade Winds that lead to the inflow of moist warm maritime air from the Indian ocean on to the continents (Fig. 13.7). It is the ascent and cooling of these monsoonal air masses on to the Himalayas that causes some of the Earth's highest, and most intense, rainfall.

Mid-latitude climates

The descent of dry (warming) air masses in the subtropical high-pressure belts leads to a general circulation there that is clockwise in the northern hemisphere and anticlockwise in the southern

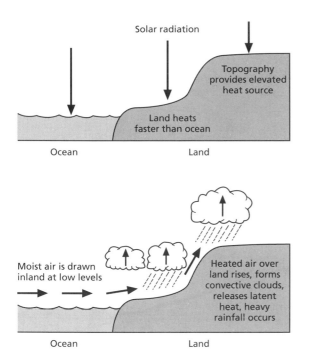

Fig. 13.7 Simplest possible representation of the monsoonal circulation. See Fig. 13.9 for more details. (After Meehl, 1992.)

the equatorial 'heat engine'. As a result, in mid-latitudes where Coriolis forces are also large, there are large radial negative gradients of pressure polewards. These pressure gradients cause high-velocity poleward flows that are turned by the strong Coriolis force to become the polar front Jet Streams. Observations indicate not only that they encircle the globe, but that the winds vary in strength and direction because of 4–5 wave-like billows that occur with wavelengths of about 2000 km or so. These are often seen as permanent features of the atmospheric circulation on mean-pressure maps. They are termed Rossby waves and serve to transfer momentum and heat across the mid-latitudes. The wavy junction between warm equatorial air and cold polar air has a large temperature gradient across it and is known as a front. The front slopes gently upwards from low to high latitudes. The intersection of the front with the surface is not simple for there are often smaller 'parasitic' waves and fronts superimposed that are shed off by the vorticity of the major Rossby waves. All these moving air masses comprise stable air of contrasting temperature and pressure separated by fronts. They dominate the weather and climate of mid-latitudes, giving rise to a more-or-less predictable sequence of weather, but at more-or-less unpredictable rates of travel.

High-latitude climates

In the polar and subpolar (> 70° latitude) regions of the Earth, short-wave insolation goes from nonexistent to low and highly seasonal. Long-wave diffuse reradiation predominates in the totally dark winters. Albedo is high all year round over permanent ice- and snowfields and high in winter everywhere because of seasonal snowfall. There is thus a net radiation deficit in high-latitude regions that is partly filled by the poleward transport of heat described above. In winter in northern latitudes there is a major radiation deficit over the Norwegian Sea associated with the formation and descent of deep water here, one of Earth's major heat sinks (see Chapter 26). Over the poles themselves there exist more-or-less permanent but weak high-pressure systems of descending air that diverge surfacewards to form the easterly polar air masses that interact with equatorial air in the mid-latitude convergence zone discussed previously. Polar skies are often cloudy and it might be thought that

hemisphere. There is also a highly important poleward-moving component that comes into contact at latitudes 30–45°N and S with equatorward-moving cool polar air, the polar easterlies. In these mid-latitude collision belts between such drastically different air masses, Coriolis deflection leads to the production of a zone of westerly winds at the circulating interface between the two air masses. This broad zone of circulation north and south of the equatorial Hadley cell is termed the *Ferrel cell*.

The westerly winds so characteristic of mid-latitude climates are really the low-altitude remnant of a much stronger wind. They extend and strengthen upwards to the top of the troposphere where very strong global westerlies, the Jet Streams, exert a most fundamental control on climate and weather. In order to understand the origins of the Jet Streams, we must go back to the equatorial story of the Hadley cell. Detailed global measurements have established that the troposphere is much thicker, particularly in summer, over low latitudes than high latitudes. This is because of strong upward motion and warming associated with these clouds should trap more reradiated energy than they actually do. The clouds are in fact very 'thin', with a sparse content of water droplets present in the very cold and undersaturated air.

13.7 Global climates: a summary

Equatorial latitudes are dominated by high insolation, low pressures, light to variable winds (these are the latitudes of the 'Doldrums'), high mean annual temperatures, low daily temperature changes, high water vapour saturation pressures and copious convective precipitation, particularly in summer months.

The dependable Trade Winds dominate latitudes 15–35°N and S of the equatorial low-pressure belt. Over land, air masses are dry and skies clear over the hot deserts, with high radiative heat transfer and consequent high daily temperature variations from very hot to cold. Over the oceans, evaporation rates into the initially unsaturated advecting air masses are very high. Hurricanes and typhoons result from instabilities set up during convective oceanic heating. The large and high landmass of southern Asia (also East Africa, South-East Asia, northern Australia and northern Mexico to lesser extents) develops its own very low-pressure system in summer that attracts and reverses the normal north-east Trade Winds, thus initiating torrential summer monsoonal precipitation.

The subtropical high-pressure belt at about latitudes 30–40°N and S is characterized by descending unsaturated air masses and generally light winds. Mediterranean climates with hot dry summers and cool wetter winters result.

The mid-latitude to temperate-latitude maritime low-pressure zones are dominated by the movement of frontal systems that form at the polar–subtropical transition. These sweep warmer saturated air north and cooler unsaturated air south as wave-like intrusions. Plentiful precipitation results in the cool to cold spring, winter and autumn seasons. Oceanic currents like the warm north-flowing Gulf Stream lead to highly important contributions to air temperatures by latent heat released in the rainstorms associated with frontal systems blowing over them. Continental areas at these latitudes (40–60°N and S) suffer much larger temperature extremes and generally lower precipitation.

The polar anticyclones preside over a stable regime of cold to very cold descending dry air masses with very high albedos over snow- and icefields under cloudless or pervasive 'thin' cloudy skies in summer giving high radiative heat losses to the atmosphere.

13.8 Climate, mountains and plateaux

It has long been appreciated that tectonic landforms have a major effect on local and regional climate. This is because they act as potential barriers that moist wind systems have to surmount or pass around (topographic 'blocking' conditions; there is an interesting analogy here between the interaction of turbidity currents with submarine obstacles, see Chapter 11). Mountain ranges also project into the troposphere and hence the local air temperatures decrease according to the value of the local lapse rate. The act of surmounting leads to reductions of both temperature and moisture, the latter since warm air can hold more water in the vapour state than cold air. This saturation effect leads to the dumping of rain on the prevailing windward slopes and summits of ranges and relatively dry conditions (the 'rain shadow') in the lee (Fig. 13.8). Since weathering and erosion rates correlate positively with available precipitation, any uplifting tectonic range will show preferential denudation on windward margins (Koons, 1989).

In recent years the role of large continental plateaux, notably the Tibetan Plateau and, to a lesser extent, the western USA, has come centre-stage in efforts to explain not only regional but also worldwide climate (see Molnar et al., 1993 for a review). Tibet in particular seems to play a major role in enhancing the Indian ocean monsoon. It is a site of high sensible heat transfer from the atmosphere in late spring, and summer and a ground-level low-pressure system results, with upward-flowing air defining an upper-atmosphere Tibetan anticyclone (Figs 13.9 & 13.10). The area thus contrasts greatly with the high-pressure Trade Wind deserts of Africa at similar latitudes, which lie in the continental interior at much lower ground elevations. A strong (> 110 km/h) easterly subtropical Jet Stream forms south of the Tibetan Plateau during the monsoon in response to strong temperature and pressure gradients between the warm rising Tibetan Plateau air and the cooler Indian ocean air. The magnitude of this vortex (formed as southward- and downward-flowing air along the Tibetan anticyclonic north to south pressure gradient is turned westwards by Coriolis force) somehow enhances the whole downward flow to contribute to the lower-level

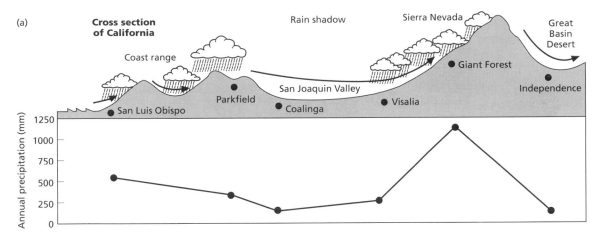

(a)

Cross section of California

Rain shadow

Sierra Nevada

Great Basin Desert

Coast range

Giant Forest

San Joaquin Valley

Independence

Parkfield

Visalia

Coalinga

San Luis Obispo

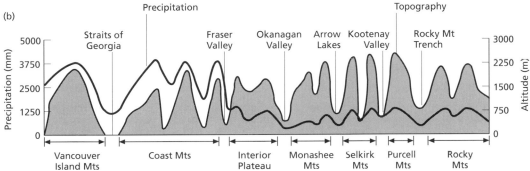

(b)

Precipitation

Topography

Straits of Georgia

Fraser Valley

Okanagan Valley

Arrow Lakes

Kootenay Valley

Rocky Mt Trench

Vancouver Island Mts

Coast Mts

Interior Plateau

Monashee Mts

Selkirk Mts

Purcell Mts

Rocky Mts

Fig. 13.8 (*above*) Elevation and precipitation profiles to illustrate orographic precipitation and rain shadow effects for (a) Sierra Nevada mountains of California and (b) the Canada–USA border. (After Barros & Lettenmaier, 1994.)

Fig. 13.9 (*below*) (a) Sketch to illustrate a 2D section through atmospheric circulations in the Indian subcontinent summer monsoon.

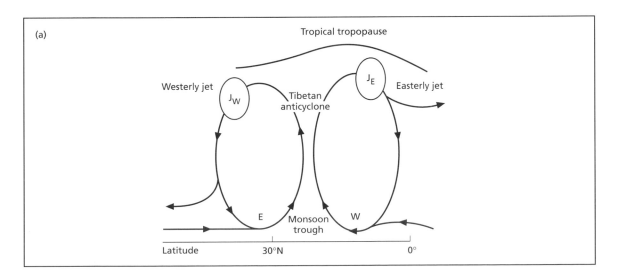

(a)

Tropical tropopause

Westerly jet

J_W

Tibetan anticyclone

J_E

Easterly jet

E

Monsoon trough

W

Latitude   30°N   0°

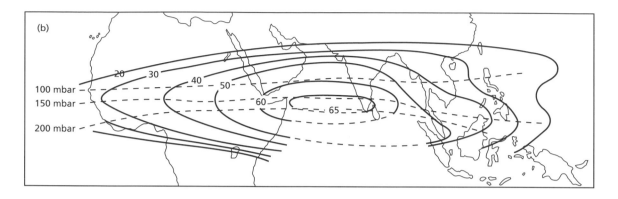

Fig. 13.9 (b) The tropical easterly jet over India; pressures in mbar and wind speed contours in knots—60 knots ≈ 110 km/h. (After Meehl, 1992.)

summer monsoonal winds, since there is a correlation between strength of the Jet Stream and the amount of monsoonal precipitation. Such a correlation is further evidence of the enhancement role of the Tibetan Plateau in the whole monsoon development. Thus the monsoonal winds from the Indian ocean (which ought to be flowing, like all Trade Winds in the northern hemisphere, to the south-west) penetrate high into the foothills (particularly the eastern foothills) and ranges of the Himalayas, contributing additional latent heat to the rising dry air on the Tibetan Plateau to the north, which is itself markedly arid during summertime.

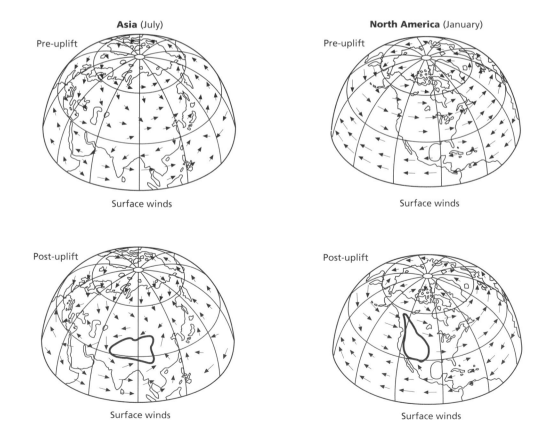

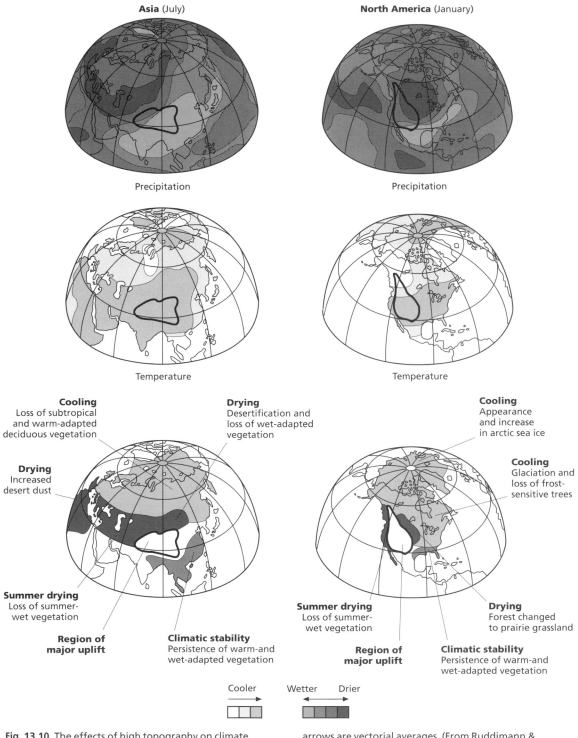

Fig. 13.10 The effects of high topography on climate in Asia and western North America. Wind speed arrows are vectorial averages. (From Ruddimann & Kutzbach, 1991.)

13.9 Climate change

We have discussed the origin and nature of today's climate, accepting the notion of a mean climate for particular regions and a net global balance between incoming and outgoing radiation. Yet looking back through recorded history (Lamb, 1995) and into the young (Wright *et al.*, 1993) and then older (Frakes *et al.*, 1992) geological record, it is obvious that significant and sometimes major changes have occurred in both regional and global climate. We shall very briefly outline the sedimentological evidence for such major climate changes below, but here it is the right time to enquire as to some of the major controls upon climate that fall out of our previous discussions. We might first enquire as to whether the global amount, surface distribution and greenhouse entrapment of incoming solar energy have remained constant through geological time. Given the great importance of the oceans in the climate machine, we must also enquire as to how known changes in continent–ocean distributions may have affected the global and regional climate.

Radiation: any change in the solar constant?

We have little direct evidence that the solar constant changes in response to short-term solar activities like sunspot cycles but there is circumstantial evidence that over periods of several hundred years the decreased activity of sunspots may be reflected in lower solar energy output since such periods are associated with severe global cooling (Little Ice Age).

Radiation: orbitally induced radiative changes?

Variations in the orbital path of Earth around the Sun and in Earth's own rotation induce longer-term (10^4–10^5 yr) changes in the *relative* solar energy flux to particular parts of the planetary surface. We stress the term 'relative' because in what follows the reader must remember that small changes in orbital parameters lead to no *net* increase or decrease in solar radiation received: the changes simply tend to apportion the radiation at different times of the solar cycle in particular hemispheres. It is this cyclical preferred apportionment that is thought to lead to climate change and the accumulation or melting of great ice-sheets. These physical changes in the seasonal distribution of incoming energy cannot of course be measured, but, in one of the great scientific breakthroughs of the 20th century, their indirect climatic effects have been carefully ascertained by sophisticated geochemical studies of Quaternary fossils.

Following the lead of the 19th-century Scottish amateur scientist James Croll, the Serbian engineer

The first orbital mechanism inducing radiation changes with time is based on the fact that Earth's rotation around the Sun is elliptical (Fig. B13.2) and not circular, a fact known since the work of Kepler in the 17th century. The very nature of an elliptical path, with the radiating Sun at one focus, means that there are seasonal variations in the amount of radiation received by Earth at aphelion and perihelion. This is based on the simple premise that the intensity of solar radiation is greatly reduced with distance from the Sun. At present Earth is nearest the Sun, by about 3×10^6 miles, on 2–3 January and furthest away on 5–6 July (note that these dates are *not* the same as the times of solstice, see below) and the solar radiation received varies by about ±3.5% from the mean value. Although these figures are not appreciable compared to the other two effects noted below, exact calculations of the gravitational effects of the other planets in the Solar System on Earth's elliptical orbit led to the later theory (due to Leverrier in 1843, I believe) of time-variable eccentricity, whereby the yearly orbit becomes more and less eccentric on the rather long timescale of 10^5 yr. At the present time we

More information on orbital changes

continued on p. 253